PUBLISHER COMMENTARY

We print NASA's handbooks and standards for the convenience of those that use them on a daily basis. We print all of these a full 8 ½ by 11 with large text so they are easy to read. Yes, color books are expensive to print so unless the information relies on the use of color for proper interpretation or understanding, we print most books in black and white to keep the cost down. All these documents are available for download for free from NASA, however printing them all over a network printer would take days.

Why buy a book you can download free? We print this so you don't have to.

All these books are available for free download from the government web site. Some are available only in electronic media. Some online docs are missing pages or barely legible.

We at 4th Watch Publishing are former government employees, so we know how government employees actually use the standards. When a new standard is released, an engineer prints it out, punches holes and puts it in a 3-ring binder. While this is not a big deal for a 5 or 10-page document, many NIST documents are over 100 pages and printing a large document is a time-consuming effort. So, an engineer that's paid $75 an hour is spending hours simply printing out the tools needed to do the job. That's time that could be better spent doing engineering. We publish these documents so engineers can focus on what they were hired to do – engineering. It's much more cost-effective to just order the latest version from Amazon.com

If there is a standard you would like published, let us know. Our web site is www.usgovpub.com

www.usgovpub.com

Copyright © 2019 4th Watch Publishing Co. All Rights Reserved

List of Other NASA Publications Available on Amazon.com:

NASA-STD-5001B	Structural Design and Test Factors of Safety for Spaceflight Hardware
NASA-STD-5006A	General Welding Requirements for Aerospace Materials
NASA-STD-5008B	Protective Coating of Carbon Steel, Stainless Steel, and Aluminum on Launch Structures, Facilities, and Ground Support Equipment
NASA-STD-5009A	Nondestructive Evaluation Requirements for Fracture-Critical Metallic Components
NASA-STD-5012B	Strength and Life Assessment Requirements for Liquid-Fueled Space Propulsion System Engines
NASA-STD-5019A	Fracture Control Requirements for Spaceflight Hardware
NASA-STD-5005D	Standard for The Design and Fabrication of Ground Support Equipment
NASA-HDBK-8739.21	Workmanship Manual for Electrostatic Discharge Control
NASA-HDBK 8739.23A	NASA Complex Electronics Handbook for Assurance Professionals (Color)
NASA-HDBK-8719.14	Handbook for Limiting Orbital Debris (Color)
NASA-HDBK-8709.22	Safety and Mission Assurance Acronyms, Abbreviations, and Definitions
NASA-HDBK-7009	NASA Handbook for Models and Simulations: An Implementation Guide For NASA-STD-7009 (Color)
NASA-HDBK-8739.19-2	Measuring and Test Equipment Specifications NASA Measurement Quality Assurance Handbook – Annex 2
NASA-HDBK-8739.19-3	Measurement Uncertainty Analysis Principles and Methods NASA Measurement Quality Assurance Handbook – Annex 3
NASA-HDBK-8739.19-4	Estimation and Evaluation of Measurement Decision Risk NASA Measurement Quality Assurance Handbook – Annex 4
NASA RCM	Reliability-Centered Maintenance Guide for Facilities and Collateral Equipment

www.usgovpub.com

Copyright © 2019 4th Watch Publishing Co. All Rights Reserved

NASA/SP-2016-3404/REV1

NASA
Work Breakdown Structure (WBS) Handbook

National Aeronautics and
Space Administration

January 2018

NASA STI Program…in Profile

Since its founding, NASA has been dedicated to the advancement of aeronautics and space science. The NASA scientific and technical information (STI) program plays a key part in helping NASA maintain this important role.

The NASA STI program operates under the auspices of the Agency Chief Information Officer. It collects, organizes, provides for archiving, and disseminates NASA's STI. The NASA STI program provides access to the NTRS Registered and its public interface, the NASA Technical Reports Server, thus providing one of the largest collections of aeronautical and space science STI in the world. Results are published in both non-NASA channels and by NASA in the NASA STI Report Series, which includes the following report types:

- TECHNICAL PUBLICATION. Reports of completed research or a major significant phase of research that present the results of NASA Programs and include extensive data or theoretical analysis. Includes compilations of significant scientific and technical data and information deemed to be of continuing reference value. NASA counter-part of peer-reviewed formal professional papers but has less stringent limitations on manuscript length and extent of graphic presentations.

- TECHNICAL MEMORANDUM. Scientific and technical findings that are preliminary or of specialized interest, e.g., quick release reports, working papers, and bibliographies that contain minimal annotation. Does not contain extensive analysis.

- CONTRACTOR REPORT. Scientific and technical findings by NASA-sponsored contractors and grantees.

- CONFERENCE PUBLICATION. Collected papers from scientific and technical conferences, symposia, seminars, or other meetings sponsored or co-sponsored by NASA.

- SPECIAL PUBLICATION. Scientific, technical, or historical information from NASA programs, projects, and missions, often concerned with subjects having substantial public interest.

- TECHNICAL TRANSLATION. English-language translations of foreign scientific and technical material pertinent to NASA's mission.

Specialized services also include organizing and publishing research results, distributing specialized research announcements and feeds, providing information desk and personal search support, and enabling data exchange services.

For more information about the NASA STI program, see the following:

- Access the NASA STI program home page at https://www.sti.nasa.gov

- E-mail your question to help@sti.nasa.gov

- Phone the NASA STI Information Desk at 757-864-9658

- Write to:
  NASA STI Information Desk
  Mail Stop 148
  NASA Langley Research Center
  Hampton, VA 23681-2199

NASA/SP-2016-3404

NASA Work Breakdown Structure (WBS) Handbook

National Aeronautics and
Space Administration

January 2018

This handbook is available in electronic form at

https://ntrs.nasa.gov

TABLE OF CONTENTS

Table of Contents	v
List of Figures and Illustrations	vii
Record of Revisions	viii
Preface	ix
P.1 Purpose	ix
P.2 Applicability	ix
P.3 References	ix
Chapter 1: Introduction	**1**
1.1 Background Information	1
1.2 Policy	1
Chapter 2: WBS Overview	**2**
2.1 Definition	2
2.2 WBS Hierarchy	3
2.2.1 Establishing and Maintaining WBS Codes in NASA's Management Systems	5
2.2.2 Contract Work Breakdown Structure (CWBS) and CWBS Dictionary	7
2.2.3 Work Breakdown Structure Elements by Other Performing Entities	9
2.3 Development Guidelines	9
2.4 Summary	10
Chapter 3: WBS Development and Control	**11**
3.1 WBS and the Project Life Cycle	11
3.2 WBS Activities and Responsibilities	12
3.3 Development Considerations	14
3.3.1 Compatibility between WBS and CWBS	14
3.3.2 Compatibility with Internal Management Systems	15
3.3.3 Correlation with Other Requirements	16
3.3.4 Number of Levels	17
3.3.5 All Inclusiveness	20
3.3.6 Change Control	21
3.4 WBS Development Techniques	21
3.4.1 Preparing Functional Requirement Block Diagrams	21
3.4.2 Coding WBS Elements in a Consistent Manner	22
3.4.3 Preparing Element Tree Diagrams	23
3.4.4 Preparing a WBS Dictionary	25
3.4.5 Using Development Checklists	28
3.4.6 Using WBS Templates	29
3.5 Common Development Errors	30
3.5.1 Using Unsuitable Former WBS	30
3.5.2 Non-Product Elements	30
3.5.3 Center Breakouts at Inappropriate Levels	31
3.5.4 Incorrect Element Hierarchy	33

Chapter 4: WBS Uses	**35**
4.1 Technical Management	36
4.1.1 Specification Tree	36
4.1.2 Configuration Management	36
4.1.3 Integrated Logistics Support	36
4.1.4 Test and Evaluation	37
4.2 Work Identification and Assignment	37
4.3 Schedule Management	38
4.4 Cost Management	39
4.5 Performance Management	40
4.6 Risk Management	41
APPENDIX A: Acronym Listing	**43**
APPENDIX B: Glossary of Terms	**45**
APPENDIX C: Standard Project WBS Level 2 Templates and WBS Dictionary Content Descriptions	**47**
APPENDIX D: Standard Data Requirements Document (DRD)	**55**
APPENDIX E: Contractor CWBS Example	**57**

List of Figures and Illustrations

2-1	Project Development Cycles and Activities..	2
2-2	WBS Levels Illustration...	4
2-3	Partial WBS with Numbering System...	5
2-4	Illustration – WBS Code Request Template for Programs/Projects...........................	7
2-5	Illustration – MdM Code Import Template...	7
2-6	WBS/CWBS Relationship..	8
3-1	WBS and the Project Life Cycle..	11
3-2	WBS Product and Enabling Support Content..	12
3-3	WBS Development Activities & Responsibilities..	14
3-4	WBS Relational Interfaces to NASA Business/Management Systems.....................	16
3-5	WBS Cross-Reference Matrix...	17
3-6	WBS Hierarchy Illustration...	18
3-7	Relationships between WBS, OBS, CA, WP, and PP..	20
3-8	Agency WBS Numbering System...	22
3-9	Partial WBS Tree Diagram Illustrating Recommended Practices.............................	24
3-10	Sample Software WBS Illustration..	25
3-11	Example – WBS Index Excerpt...	26
3-12	WBS Dictionary Example..	28
3-13	WBS Checklist Example..	29
3-14	Unsuitable Non-Product, Phase-Oriented WBS...	31
3-15	Unsuitable Functional/Organizational Oriented WBS...	31
3-16	Center Breakout Guidance for a WBS...	33
3-17	Illustration of Incorrect Element Hierarchy..	34
4-1	The WBS as a Project Management Tool for Integration..	35
4-2	Responsibility Assignment Matrix (RAM)...	38
4-3	WBS and the Development of the Performance Measurement Baseline..................	41
4-4	WBS Serves as a Common Reference Point in Risk Management..........................	42

Record of Revisions

REV LTR	DESCRIPTION	DATE
	Basic Issue	January 2010
A	Miscellaneous Minor Revisions	October 2016
B	Miscellaneous Minor Revisions	January 2018

Preface

P.1 Purpose

The purpose of this document is to provide program/project teams necessary instruction and guidance in the best practices for Work Breakdown Structure (WBS) and WBS dictionary development and use for project implementation and management control. This handbook can be used for all types of NASA projects and work activities including research, development, construction, test and evaluation, and operations. The products of these work efforts may be hardware, software, data, or service elements (alone or in combination). The aim of this document is to assist project teams in the development of effective work breakdown structures that provide a framework of common reference for all project elements.

The WBS and WBS dictionary are effective management processes for planning, organizing, and administering NASA programs and projects. The guidance contained in this document is applicable to both in-house, NASA-led effort and contracted effort. It assists management teams from both entities in fulfilling necessary responsibilities for successful accomplishment of project cost, schedule, and technical goals.

Benefits resulting from the use of an effective WBS include, but are not limited to: providing a basis for assigned project responsibilities, providing a basis for project schedule and budget development, simplifying a project by dividing the total work scope into manageable units, and providing a common reference for all project communication.

P.2 Applicability

This handbook provides WBS and WBS dictionary development guidance for NASA Headquarters, NASA Centers, the Jet Propulsion Laboratory, inter-government partners, academic institutions, international partners, and contractors to the extent specified in the contract or agreement.

P.3 References

NPD 7120.4, *NASA Engineering and Program/Project Management Policy*
NFS Part 1834, *Major Systems Acquisition*
Electronic Industries Alliance (EIA)-748, *Earned Value Management Systems Standard*
NPR 7120.5, *NASA Space Flight Program and Project Management Requirements*
NPR 7120.7, *NASA Information Technology and Institutional Infrastructure Program and Project Requirements*
NPR 7120.8, *NASA Research and Technology Program and Project Management Requirements*
MIL-STD-881, *Department of Defense Standard Practice, Work Breakdown Structures for Defense Materiel Items*
PMI 978-1-933890-13-5, *Practice Standard for Work Breakdown Structures*
NASA/SP-3705, *NASA Space Flight Program and Project Management Handbook*
NASA/SP-6105, *NASA Systems Engineering Handbook*

Chapter 1: Introduction

1.1   Background Information

In accordance with NFS Part 1834, *Major Systems Acquisition,* NASA policy NPD 7120.4, *NASA Engineering and Program/Project Management Policy, and* directives NPR 7120.5, *NASA Space Flight Program and Project Management Requirements*, NPR 7120.7, *NASA Information Technology and Institutional Infrastructure Program and Project Requirements (NID 7120.99 Interim Directive),* NPR 7120.8, *NASA Research and Technology Program and Project Management Requirements,* the WBS and WBS Dictionary are mandatory elements of a project's management baseline. This section provides general WBS information including policy, definition, guidelines, and development process.

1.2   Policy

Per NPR 7120.5, NPR 7120.7/NID 7120.99 Interim Directive, NPR 7120.8, and NFS Part 1834 a project WBS is a key element of NASA project management processes. The WBS and WBS Dictionary requirements contained in these three documents apply to all types of NASA programs and projects depending on the product line involved. The WBS is a core element of a project's baseline throughout all life cycle phases. It is the responsibility of each project manager and their project team to ensure that the WBS requirements are adhered to, not only during initial WBS development, but also in its on-going maintenance and control. The standard project WBS structures and templates identified in the above NPRs were intended to apply only to new projects established on or after June 1, 2005.

Chapter 2: WBS Overview

2.1 Definition

Each NASA program has a set of goals which are developed from NASA mission needs. These program goals are expanded into specific project objectives. The function of management is to plan and direct project activities to achieve the program goals. The WBS is to be generated and utilized as a key tool by management in order to execute the project/program.

A WBS is a product-oriented family tree that identifies the hardware, software, services, and all other deliverables required to achieve an end project objective. The purpose of a WBS is to subdivide the project's work content into manageable segments to facilitate planning and control of cost, schedule, and technical content. A WBS is developed early in the project development cycle as reflected below within Figure 2-1. It identifies the total project work to be performed, which includes not only all NASA in-house work content, but also all work content to be performed by contractors, international partners, universities, or any other performing entities. Work scope not contained in the project WBS should not be considered part of the project. The WBS divides the work content into manageable elements, with increasing levels of detail.

INITIATING	PLANNING	EXECUTING	CONTROLLING	CLOSING
• Establish Project Objectives - Program Formulation Authorization - Program Commitment Agreement - Program Plan - Project Formulation Approval - Operations Concept Development (Design Reference Missions)	• Define the Work - Requirements - Functional/Physical/Operational Architecture - System Interfaces - **Work Breakdown Structure** - Identify Risks - Contract RFP - Project Plan • Schedule Work and Resources - Develop Resource Loaded Schedule - Determine Project Costs and Margins - Define Responsibility Matrix	• Perform the Work - Design - Review - Fabricate - Test - Operate - Complete • Provide Reports - Health Status - Earned Value - Management and Independent Reviews (including PMC) - Risk Management	• Track Actual Performance • Analyze Project Progress • Initiate Corrective Action • Replan as Required	**Complete the Project**

Figure 2-1: Project Development Cycles and Activities

A WBS is developed by first identifying the system or project end item to be structured, and then successively subdividing it into increasingly detailed and manageable subsidiary work products or elements. Most of these elements are the direct result of work (e.g., assemblies, subassemblies, and components), while others are simply the aggregation of selected products into logical sets (e.g., buildings and utilities) for management control purposes. In either case, the subsidiary work product has its own set of goals and objectives which must be met in order for the project objectives to be met. Detailed tasks which must be performed to satisfy the subsidiary work product goals and objectives are then identified and defined for each work product or element on which work will be performed.

Completion of an element is both measurable and verifiable based upon specific completion criteria established during upfront project planning by the project team. Because WBS element/product completion can be verified, a WBS provides a solid basis for technical, schedule, and cost plans and status. No other structure (e.g., code of account, functional organization, budget and reporting, cost element) satisfactorily provides an equally solid basis for incremental project performance assessment.

2.2 WBS Hierarchy

The project WBS structure should encompass the entire project's approved scope of work. It usually consists of multiple levels of products along with associated work content definitions that are contained in a companion document called the WBS Dictionary. All NASA projects have the capability of subdividing the work content down to any level necessary for management and insight. However, the Agency's Core Financial System currently limits the ability to capture costs to a maximum of seven levels. These seven levels of the WBS are defined below.

- Level 1 is the entire project.
- Level 2 elements are the major operational product elements along with key common, enabling products (as defined in NPR 7120.5, NPR 7120.7 (NID 7120.99 Interim Directive), and NPR 7120.8 standard WBS templates).
- Level 3-7 contains further definable subdivisions of the products contained in the level 2 elements (e.g., subsystems, components, documents, functionality).

There are numerous terms used to define level three and succeeding levels of the WBS below the system level. Some typical examples used for hardware and software product elements are subsystem, subassembly, component, module, functionality, equipment, and part. Project management and other enabling organizational support products should use the subdivisions and terms that most effectively and accurately depict the hierarchical breakdown of project work into meaningful products.

A properly structured WBS will readily allow complete aggregation of cost, schedule, and performance data from lower elements up to the project or program level without allocation of a single element of work scope to two or more WBS elements. WBS elements should be identified by a clear, descriptive title and by a numbering scheme as defined by the project that performs the following functions:

- Identifies the level of the WBS element.
- Identifies the higher-level element into which the element will be integrated.

The following general illustration depicts how work scope can be arranged as hierarchical WBS levels of work within a project. All project effort must be included, including all NASA in-house, contracted, international partner, university, and any other performing entity implementations. Enabling organizational common products must also be reflected appropriately with a project WBS (e.g., Project Management, Safety & Management Assurance (S&MA), Systems Engineering and Integration (SE&I)).

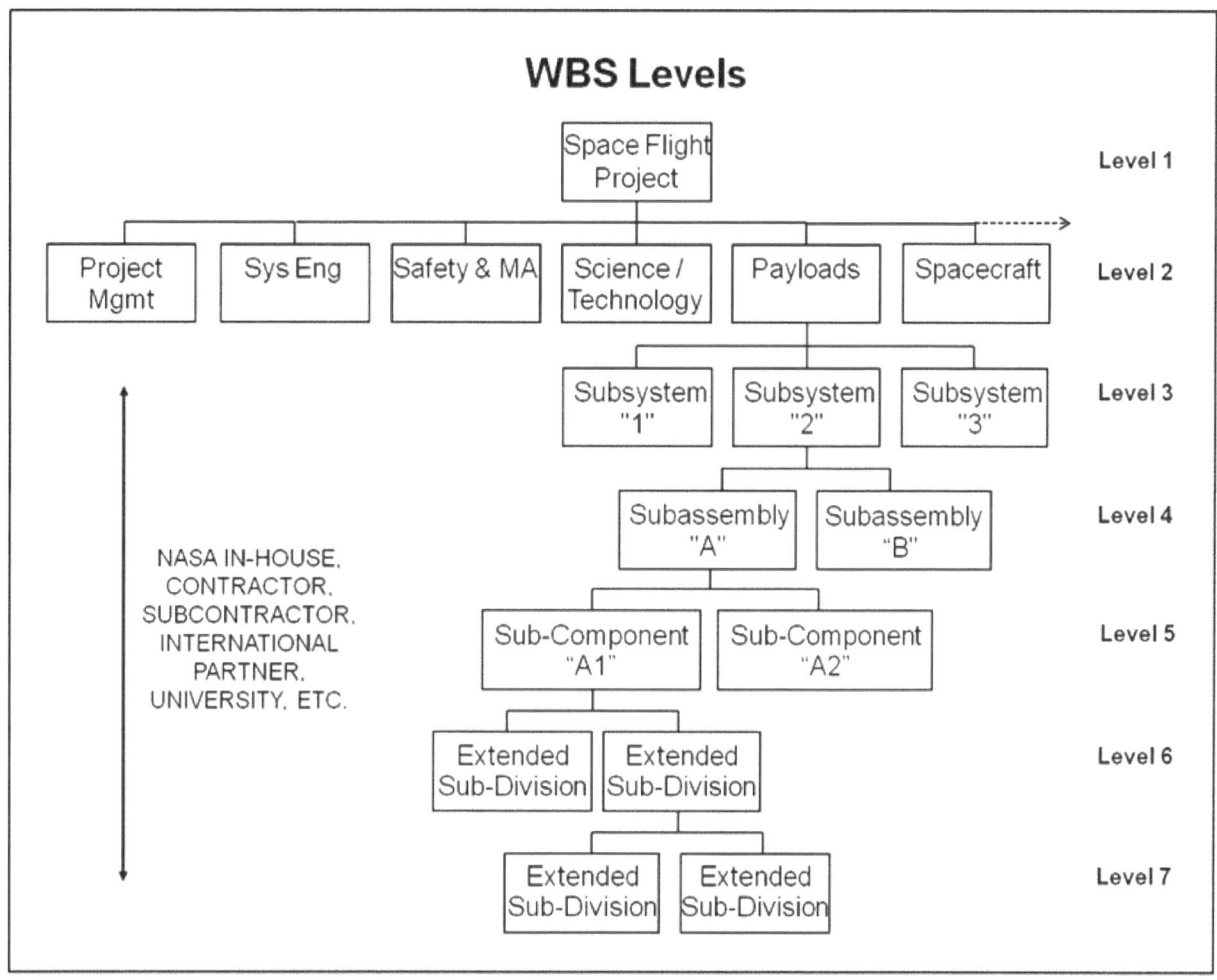

Figure 2-2: WBS Levels Illustration

The following portion of a project WBS reflects an example of the NASA authorized WBS numbering system. This numbering scheme is called the NASA Structure Management (NSM) system. For each Agency project, the WBS established by the project team must use the NSM numbering scheme and also must correlate exactly through level seven to the corresponding financial accounting structure utilized for each project within the NASA Core Financial System. This requirement helps to ensure that project costs are applied to the correct work scope being implemented by the project. This process is necessary for carrying out successful Earned Value Management (EVM) processes.

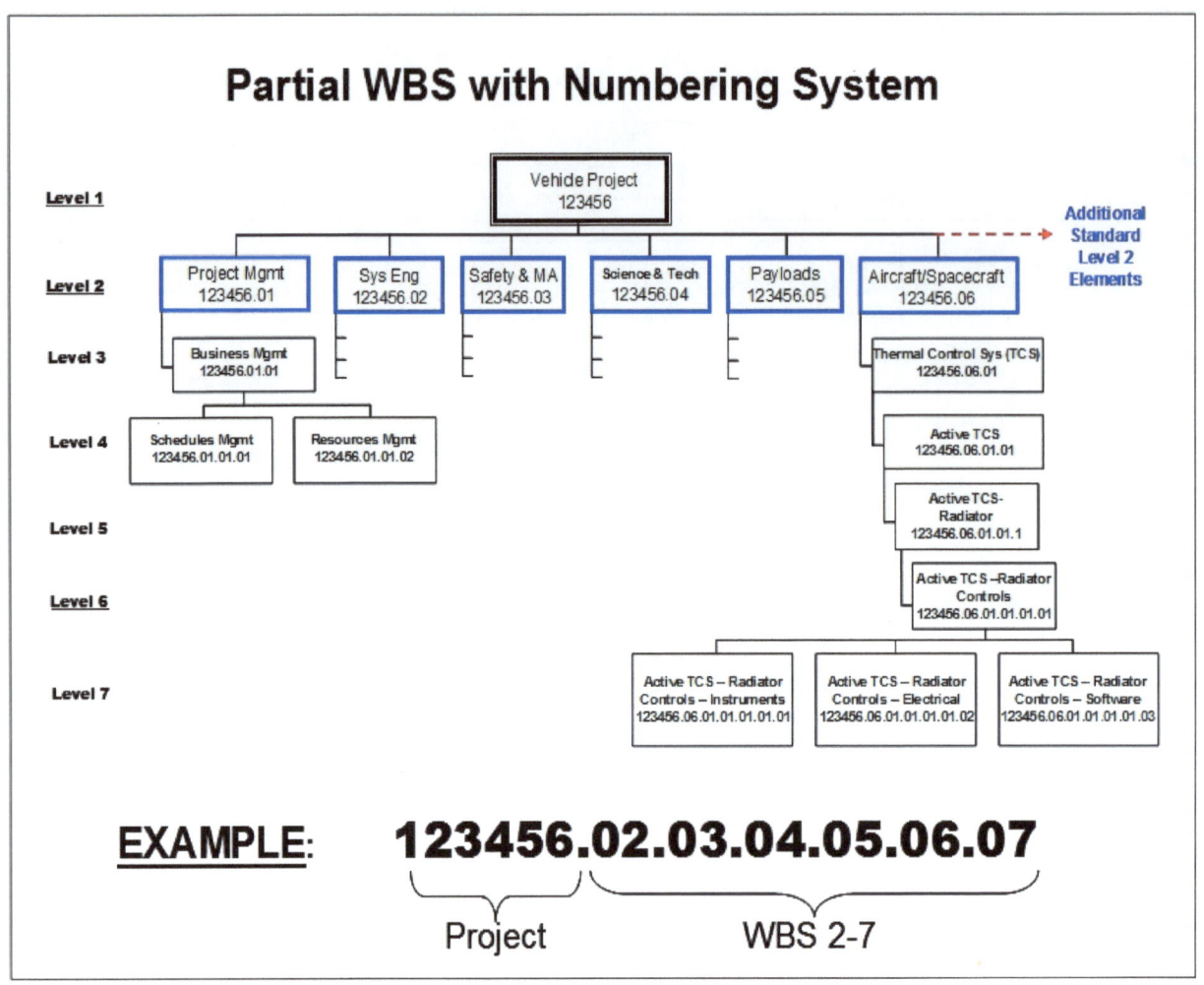

Figure 2-3: Partial WBS with Numbering System

The top two levels of a project WBS are dictated and controlled by the Agency through standard, level-two WBS templates. These templates, along with their associated narrative content descriptions, are contained in the NASA Space Flight Program and Project Management Handbook (NASA/SP-3705), NPR 7120.7 (NID 7120.99 Interim Directive Appendix H for Information Technology projects), and NPR 7120.8 (Appendix K for Technology Development projects). WBS levels 3 and lower are developed and should be controlled by project management and, as-required, prime contractors that are involved in project implementation. In cases where prime contractors are involved, lower-level element coding must be traceable to the appropriate upper-level elements that are controlled by the NASA Project Manager. While not being a requirement, it is recommended that the prime contractor lower-level WBS numbering scheme be consistent with the overall project WBS numbering format. This will allow easier total project integration of cost and EVM data for project reporting.

NASA standard level-two WBS templates and narrative descriptions can be found in Appendix C.

2.2.1 Establishing and Maintaining WBS Codes in NASA's Management Systems

All Programmatic and Institutional WBS element codes are not recognized as official NASA structures until first being approved and established in the Agency's Metadata Management (MdM) system. The

MdM system is a web-based enterprise application that contains the Agency's official NSM data elements and associated attributes. MdM is the only Agency application used for identifying, creating, tracking, organizing and archiving of Appropriation, Mission, Theme, Program, Project, and Work Breakdown Structure (WBS) 2 through 7 NSM structural elements. As the Agency's enterprise repository for NSM data, MdM supplies WBS codes to the Agency's Core Financial System and the Budget Formulation System as they require programmatic and hierarchical coding of data content. The WBS approval process involves designated MdM code requesters/approvers that have been established across the Agency to review new WBS elements requested by programmatic and institutional organizations. It should be noted that project managers are not currently included as MdM code approvers. Because of this, all project managers should continually monitor new WBS elements that are added to their projects for validity and correctness.

Process instructions for entering new or modifying existing, WBS elements within the MdM system may be obtained from the designated MdM code requester point of contact at each NASA Center. Additional information regarding the MdM system may also be obtained by contacting the MdM Help Desk (mdmhelpdesk@hq.nasa.gov). All modifications made to existing WBS element codes contained in Agency management systems listed above must also first be initiated and approved through the MdM System. A WBS code that has been approved and officially entered into the MdM System cannot be removed. This restriction enhances a project's ability to maintain accurate historical project data.

As a program/project or institutional organization determines the need, a code request may be submitted to the authorized Center MdM code requester that addresses any of the following MdM activity categories:

- The creation of new WBS elements.
- The modification of attribute data associated with any WBS element.
- The total closure of a WBS element so that it is unavailable for any further Commitment, Obligation, Cost, and Disbursement (COCD).
- The "technical" closure a WBS element so that it is unavailable for any further Commitments and Obligations, but does allow availability for any final costs or disbursements for the element.
- The "retirement" of WBS elements that are still in the Formulation structure and haven't been approved to receive funds.

The MdM code request involves the use of standard data templates. Figure 2-4 reflects an illustration of just one example of a WBS code request template that may be used by program/project teams or NASA institutional organizations. It should be understood that each NASA center will potentially have a different code request template for use in submitting requests for new or modified WBS elements to their authorized MdM code request point of contact. The center MdM code request point of contact will then use an MdM import template (see Figure 2-5) to prepare requested data in the proper format for review and import by the responsible MdM system office at NASA Headquarters. The request templates aid in ensuring that all needed data is included and that the proper formatting is followed to enable each request to be adequately reviewed, approved, and entered into the MdM system by the authorized code requester.

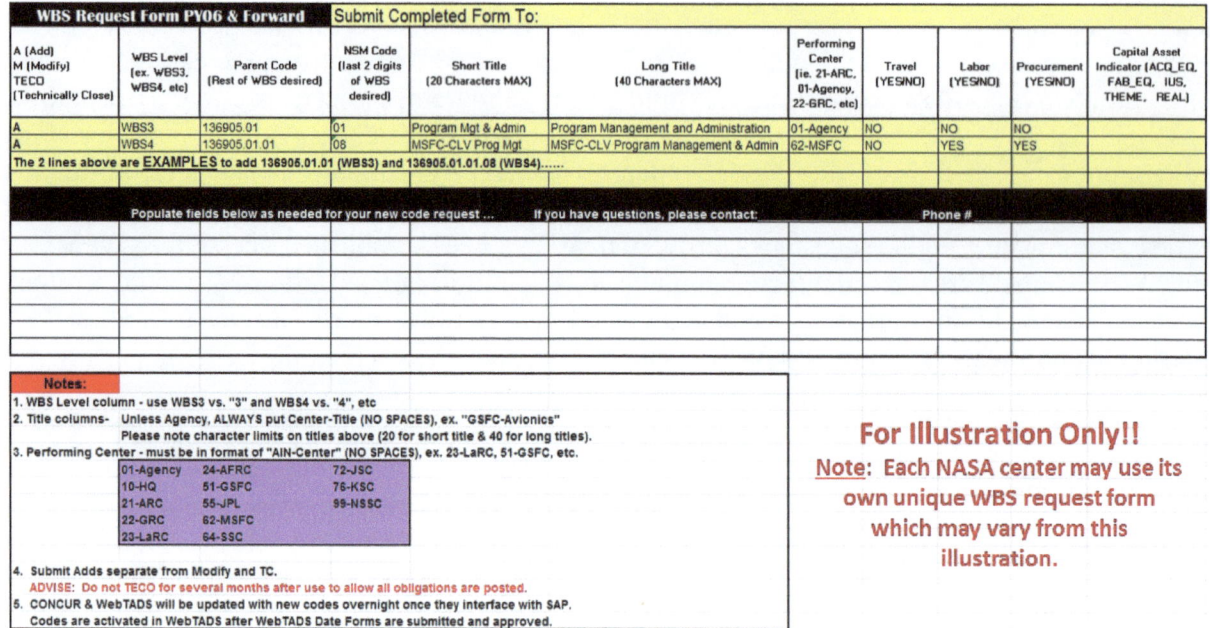

Figure 2-4: Illustration – WBS Code Request Template for Programs/Projects

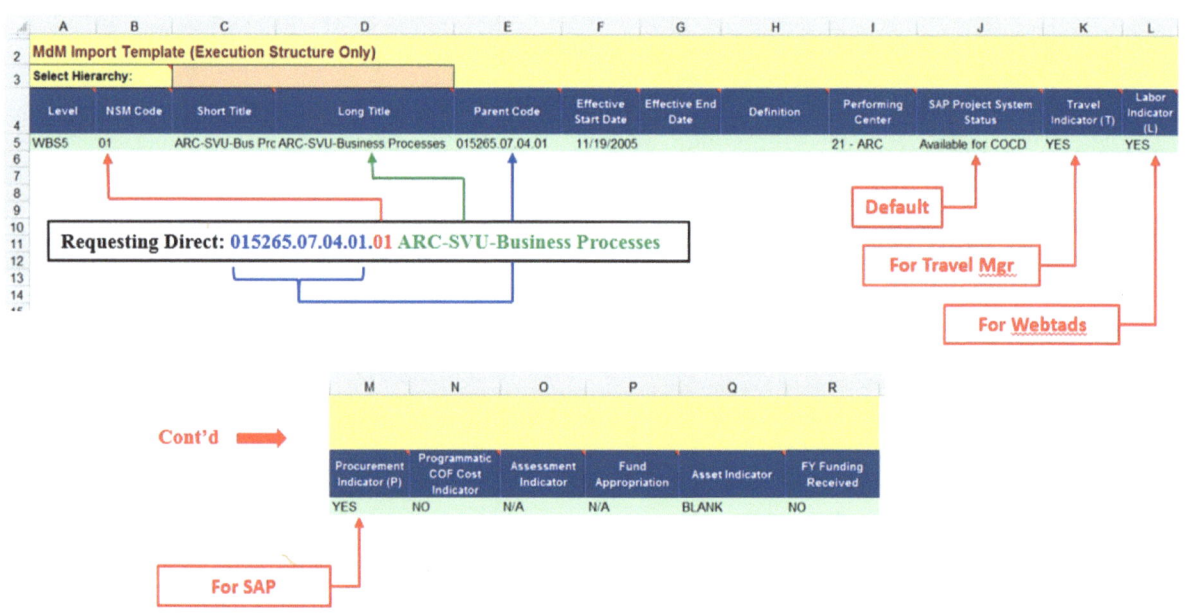

Figure 2-5: Illustration – MdM Code Import Template

2.2.2 Contract Work Breakdown Structure (CWBS) and CWBS Dictionary

For projects involving significant contracted effort, a project's preliminary WBS should be included in the Request for Proposal (RFP) and used as a starting point for individual contractors to develop their extended CWBS. It is important to remember that each project will have only one WBS and that the contractor's CWBS is just an extension of the upper-level project WBS elements. The RFP will also contain specific instructions to contractors that their proposal should be submitted in accordance with the specified preliminary WBS. As noted in the previous section, the contractor's WBS coding scheme must be traceable to upper-level project controlled elements. Again, while not a requirement, it is highly recommended that the extended CWBS be developed using a numbering format that is consistent with or relatable to the standard NASA project WBS element numbering format.

The CWBS contains the complete WBS hierarchy for a specific contract scope of work. It is developed by the contractor in accordance with the contract statement of work (SOW). It includes the WBS elements for the products which are to be furnished by the contractor. The contractor extends these elements and defines the lower-level products. It should be noted that within the contractor's own internal management systems and WBS process, there may not be the same limitation of a maximum of seven levels within the product hierarchy, as is the case for NASA's WBS hierarchical format. However, as NASA projects reflect a total integrated WBS structure (containing both in-house and contractor effort), only the upper-element levels of contractor products can typically be included due to the seven-level limitation of the NSM numbering system. The contractor reporting requirements will indicate the CWBS levels or elements for which contract status is to be reported to the government. Contractor WBS reporting requirements are contained in a standard contractor Data Requirements Document (DRD) which should be a part of the RFP. A standard DRD template is shown in Appendix D.

A properly formulated CWBS provides a consistent and visible framework that facilitates uniform planning, assignment of responsibilities, data summarization, and status reporting. A properly formulated CWBS is also a key consideration in the successful implementation of EVM processes.

The following figure illustrates the relationship of an overall project WBS and its CWBS elements:

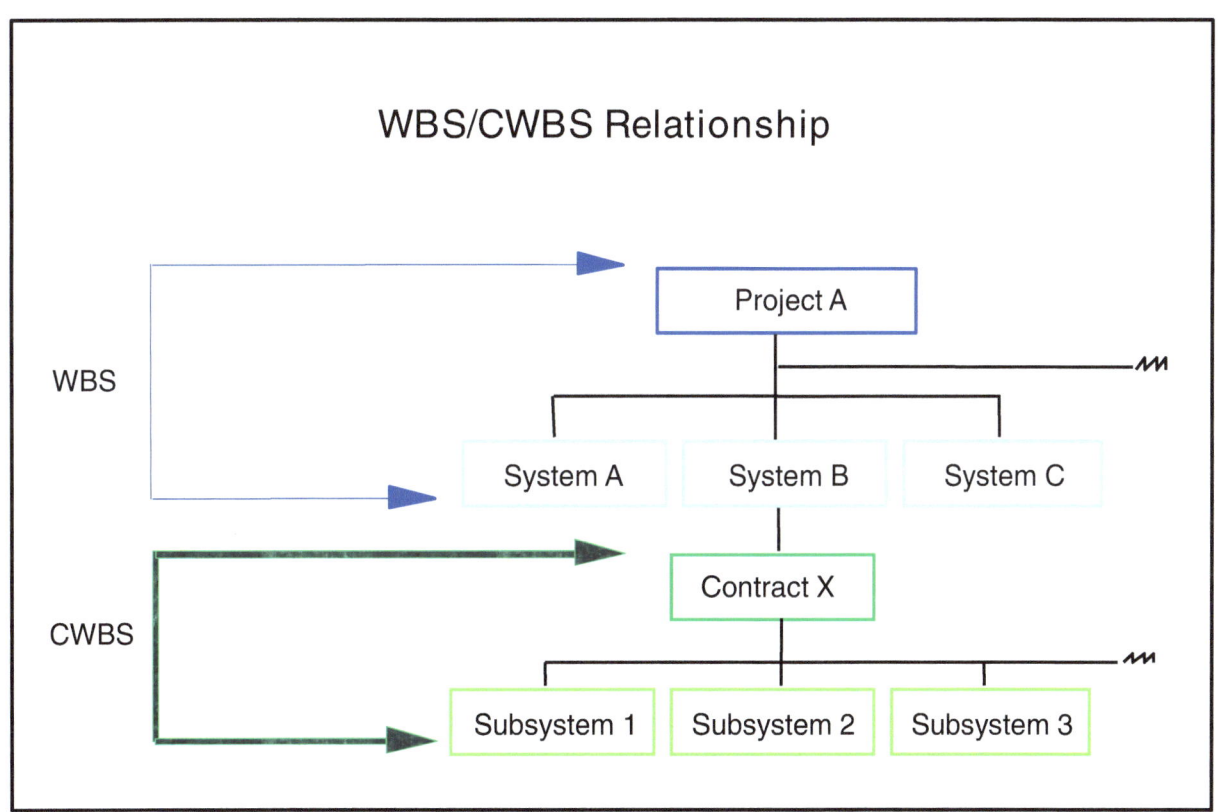

Figure 2-6: WBS/CWBS Relationship

The following is a typical contract clause used for incorporating the CWBS into a contract. Project managers should work with the contracts or procurement organization to develop the desired contractual language for such a clause.

> "A Contract Work Breakdown Structure has been negotiated between the government and the contractor. The top levels of the Contract Work Breakdown Structure are formally incorporated into the contract as set forth in Appendix E. The elements shown in this exhibit may not be changed except by contractual action. Lower-tier elements which are not shown in this exhibit may be changed by the contractor as appropriate, provided that notification of such changes is provided to NASA's Contracting Officer."

A CWBS example is found in Appendix E.

2.2.3 Work Breakdown Structure Elements by Other Performing Entities

For projects involving scope content being implemented by other performing entities, such as *(but not limited to)* international partners and universities, the WBS should also reflect this work content within the overall hierarchy of project work. While other performing entities may, or may not have standard contractual arrangements, they are nonetheless responsible for specified WBS elements through some type of directed agreement arrangement with NASA. This work content must also be subdivided to an appropriate level of product-oriented detail for project planning, control, and reporting. The resulting work elements must be clearly identified and included within the project WBS under the correct hierarchical branches in just the same manner as contractor CWBS elements. Again, while not a requirement, it is highly recommended that the extended WBS for other non-NASA work scope be developed using a numbering format that is consistent with or relatable to the standard NASA WBS element numbering format, or at a minimum clearly traceable to the correct upper-level WBS hierarchical elements.

2.3 Development Guidelines

Only one WBS is prepared for each NASA project and includes both in-house and contractor effort. While there is no single "right way" to prepare and utilize a WBS, there are some generally recommended guidelines that should be followed. Considering the following general guidance will assist in creating and implementing a WBS.

a. The WBS is prepared as early as project definition will permit.
b. A preliminary WBS is initially developed early in project formulation to define the top levels of a WBS. These preliminary elements should reflect the entire scope of work contained in the overall project life cycle including project definition, development, launch, and operations.
c. A single WBS is used for both technical and business management through level seven.
d. Both high level and detailed WBS planning should involve all stakeholders to ensure that proper planning is done and that ALL parties agree on the final WBS prior to approval.
e. When a project is authorized by a Program Commitment Agreement (PCA), the WBS becomes formalized as the project outline; changes to it must be formally approved by the program office.
f. A preliminary CWBS is developed from the basic elements of the preliminary WBS and expanded for use in the RFP, preparation of proposals, and the evaluation and selection process.
g. Normally, only the second or third level elements of the preliminary CWBS will be specified by NASA in an RFP. The CWBS is considered a preliminary CWBS until it is finalized as a result of negotiation and incorporated formally into the contract.

h. A total project WBS is created by combining the elements of the NASA in-house WBS elements with the contractor's CWBS elements, and also by all other WBS elements being implemented by other performing entities.
i. As the project scope of work changes, the WBS is revised to reflect changes that are formally approved through the configuration control process.
j. All top-level WBS elements do not have to be sub-divided to the same level of detail. As associated element risk, cost, and/or complexity increases, further breakdown may be necessary.
k. When high-risk items are located at low CWBS levels *(also low WBS levels by other performing entities)*, these items can be identified against the higher-level WBS or CWBS element of which the high risk item is a part.

While most project WBS structures are different at lower levels, the above guidance can help guarantee that level one and two WBS elements are consistent across NASA projects. This ensures proper and recommended WBS characteristics exist for all projects.

2.4 Summary

As previously discussed, a WBS defines all work to be performed for project completion. It is a product-oriented structure, not an organizational structure. To develop and maintain a WBS, you must have a clear understanding of the project's objectives and the end item(s) or end product(s) of the work to be performed. The WBS provides a common reference for all project communication, both internally within the team and externally to project stakeholders. The WBS also provides a means of rolling up project data to any desired level for analysis and oversight.

Because of its product orientation, a WBS provides the framework to plan, track and assess the project's technical, schedule, and cost performance.

Chapter 3: WBS Development and Control

3.1 WBS and the Project Life Cycle

A preliminary WBS is developed early in the conceptual stages of the project. It is established as soon as program management believes the project has reached a stage of definition where it is feasible. It is used to assist in the preparation of the PCA and the project plan. The preliminary project planning process is an iterative process. During its early phases, the preliminary WBS may be revised as necessary. Once the project is established with sufficient and stable scope definition, then both the NASA in-house work elements and, if required, the CWBS elements can be adequately planned and established at the necessary levels of detail.

CAUTION: Do not enter the full preliminary WBS into MdM system until the project's definition of work content has stabilized or until specific charge codes are needed for cost collection. The chart below (Figure 3-1) reflects how the WBS development process relates to the project life cycle.

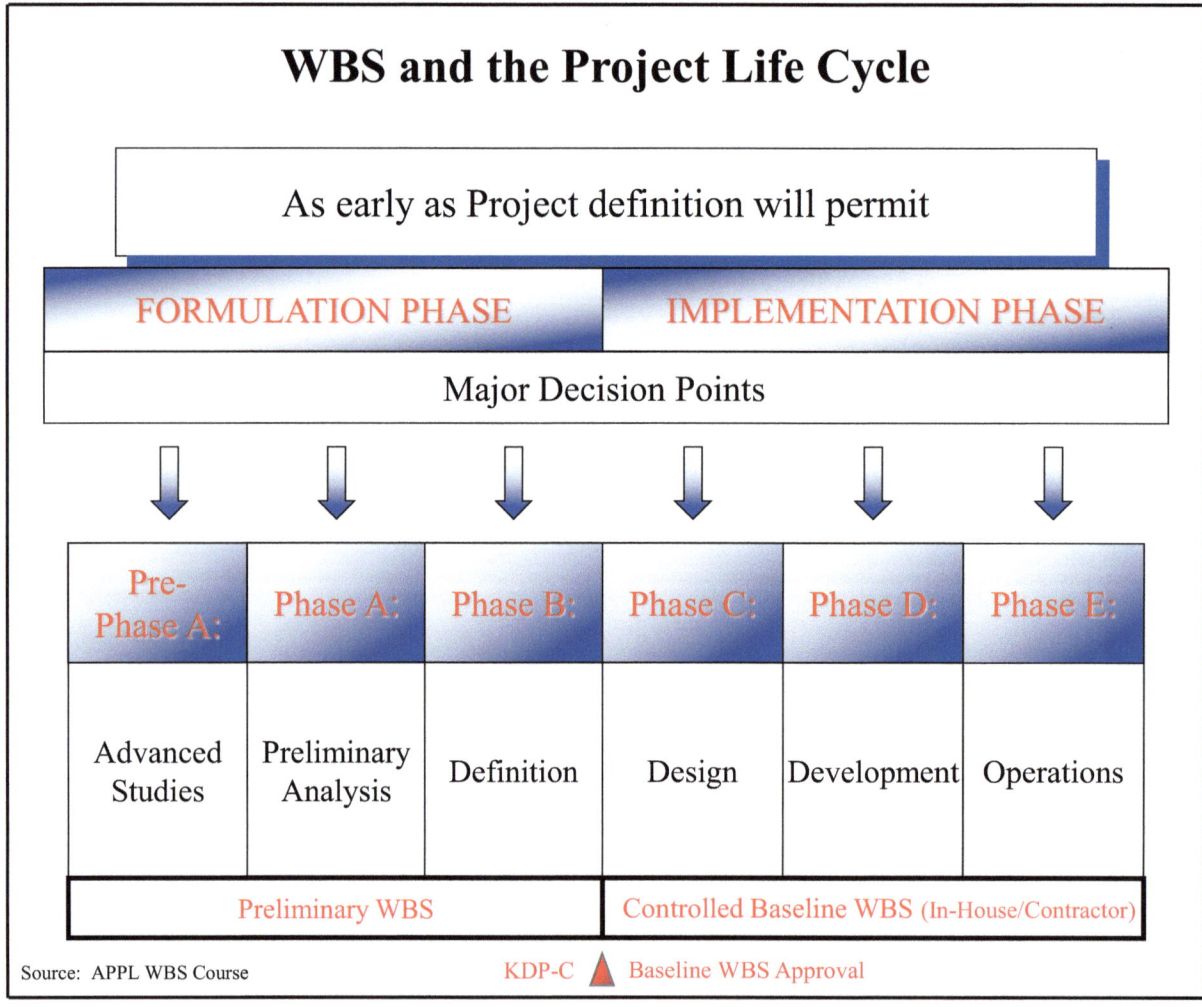

Figure 3-1: WBS and the Project Life Cycle

Baseline approval of the WBS is achieved during the latter stages of project life-cycle phase "B". Once approved, the WBS should not be revised except through the formal approval process at major transitional points in the project.

3.2 WBS Activities and Responsibilities

The project manager is responsible for developing and maintaining the WBS and the WBS Dictionary. This responsibility applies to WBS structures that are predominantly in-house NASA implemented effort, contracted efforts requiring negotiation and approval of CWBS elements, and also to WBS elements that are implemented by other performing entities.

The initial preliminary WBS may be developed as a totally new document based only on the understanding of work scope at that time. Or a new WBS may also be developed from an existing structure from a past similar project. The WBS evolves from an iterative analysis of:

a. Project objectives
b. Functional design criteria
c. Project scope
d. Technical performance requirements
e. Proposed methods of performance
f. Drawings, process flow charts, and other technical documentation
g. Risk and complexity (Note: Cost and schedule changes may impact scope which could drive modifications to the WBS.)

As stated earlier, a project WBS, while being predominantly product-oriented, must also reflect all the effort that is in the approved scope of work. Because of this, common elements of work must be included that are not necessarily product-oriented. This effort is typically accomplished by "enabling organizations," such as project management, safety and mission assurance, etc. The chart below (Figure 3-2) reflects how a project WBS contains both product-oriented elements and common elements performed by enabling organizational support.

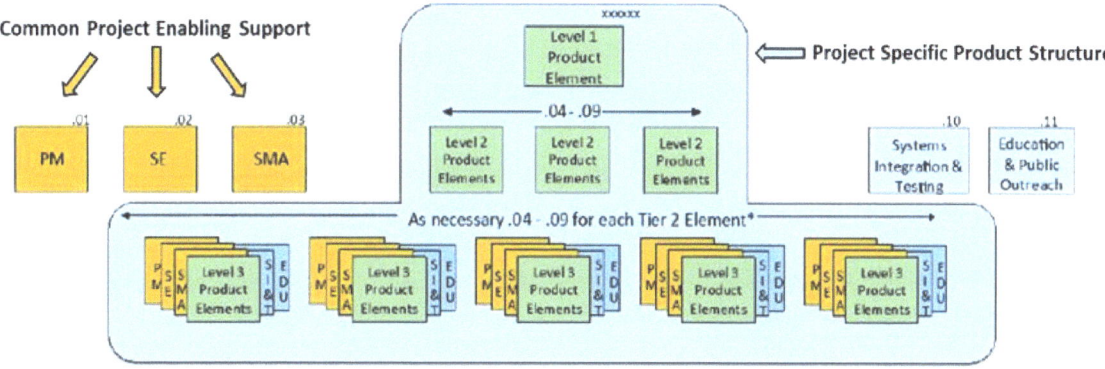

Figure 3-2: WBS Product and Enabling Support Content

The WBS should be developed by the project team and approved by the project manager. This development must be a coordinated effort involving the following participants:

a. Systems Engineering
b. Project Management

c. Business Management
   d. Project Planning and Control (PP&C) Disciplines
   e. Technical Disciplines
   f. Safety & Mission Assurance

Projects with scopes of work to be implemented by predominantly in-house civil servants and associated support contractors must also reflect their total effort with appropriate hierarchical subdivided WBS elements. This subdivision of work must extend down to the point where work is actually accomplished and charged.

The NASA project office incorporates a preliminary WBS into each RFP by selecting the appropriate WBS elements for the work products that will be required by each contract. The WBS should be accompanied by the initial WBS dictionary, which is a narrative definition of each element appearing on the WBS. The RFP instructs potential contractors to extend the selected CWBS elements appropriately.

Once NASA issues an RFP, the contractor extends the selected CWBS elements appropriately to create a proposed CWBS to be submitted with the proposal. The RFP should also indicate the specified format and content desired for the WBS Dictionary.

Contractors may suggest changes to selected CWBS elements when needed in order to meet an essential requirement of the RFP or to enhance the effectiveness of the CWBS in satisfying the project objectives. In preparing a contract proposal, a contractor may determine that the RFP requirements pertaining to the CWBS would force some unusual requirements on existing contractor management control systems. If this is the case, the contractor may request modifications to specific CWBS requirements to minimize the impact on the contractor's existing management control systems.

As part of the proposal evaluation, a NASA team performs a technical and programmatic evaluation of the CWBS submitted by each proposer. A contractor's proposal may occasionally include alternate technical approaches that impact the preliminary CWBS content contained in the RFP. In these cases, NASA management must carefully evaluate the proposed CWBS changes and ensure the project content is adequately addressed to successfully accomplish the contract objectives. Fortunately, the procurement process provides specific opportunities for discussion and resolution of issues identified within the proposal. Therefore, it is crucial that during these discussions, any exceptions taken by NASA to a contractor's proposed preliminary CWBS or CWBS Dictionary should be pointed out to the contractor, so that an acceptable CWBS will be included with the contractor's final proposal revision and the resulting contract.

The contractor maintains the extended CWBS and CWBS Dictionary, including change traceability. Only NASA-approved changes may be incorporated in accordance with the contract terms.

The following table (Figure 3-3) summarizes the WBS activities and identifies the responsible party for each activity.

WBS Activity	Responsible Party	
	Project Manager	Contractor or Other Performing Entity
Prepare preliminary WBS and WBS dictionary	X	
Select WBS elements for CWBS and include in RFP (if acquisition required)	X	
Select WBS elements for other non-NASA performing entities (if required)	X	
Extend (decompose) WBS for in-house efforts to levels where costs will be collected	X	
Submit WBS structure and elements only if necessary to MdM system for cost collection (during early Phase A & B effort)	X	
Select WBS elements required for early phase cost collection and activation of charge numbers	X	
Extend (decompose) CWBS and/or other non-NASA performing entities WBS to necessary levels and submit in proposal		X
In-house project team evaluate and approve WBS & WBS Dictionary	X	
NASA project team to conduct evaluation of proposed CWBS of other non-NASA performing entities WBS (if applicable)	X	
Negotiate & update proposed CWBS & CWBS Dictionary and/or Other Performing Entity WBS & WBS Dictionary (if applicable)	X	X
Submit approved project WBS and CWBS to HQ's MdM system	X	
Select additional appropriate WBS elements for cost collection and activation of charge numbers	X	
Maintain baseline WBS & WBS Dictionary (include CWBS & CWBS Dictionary)	X	X

Figure 3-3: WBS Development Activities & Responsibilities

3.3 Development Considerations

These items should be considered when developing a WBS or CWBS:

a. Compatibility between WBS and CWBS
b. Compatibility with internal management systems
c. Consistency with NASA Core Financial System
d. Correlation with other technical or programmatic requirements
e. Number of levels
f. All inclusiveness
g. Change control
h. Control Account (CA) levels for charging costs (for in-house NASA effort)

Following is a brief discussion of each consideration. Checklists to aid in the development process are found in section 3.4.5 of Development Techniques.

3.3.1 Compatibility between WBS and CWBS

Each CWBS must be an extension of the WBS elements selected by NASA, and must be structured and coded so that technical, schedule, and cost information may be readily summarized into the WBS. There should be no work content included in any CWBS element that falls outside the defined work content of the associated upper-level WBS element. In turn, the WBS must accommodate the management and business systems' needs of the winning contractors to a reasonable and practical extent.

In NASA's current development and management environment, it is not unusual for projects to establish a single extensive contract where the scope of work includes product content that crosses multiple WBS elements. One example of this situation may involve a prime contractor with an approved scope of work that entails development of hardware and/or software products for multiple WBS system elements. Another example may involve a contract with a broad scope of work to supply support contractors to assist government personnel in project work that leads to products in the hierarchies of multiple WBS elements. In both of these cases, it is recommended that the contract scope of work be subdivided in a product-oriented manner that allows for the most accurate depiction of the hierarchical breakdown of project work into meaningful products. This means that a single contract scope of work may entail multiple, clearly identifiable work products that require their hierarchical positioning under multiple upper-level WBS elements. Proper structuring in this manner will readily allow a more complete and meaningful aggregation of cost, schedule, and performance data for management use. Please note that in these cases, it is imperative that the contracts are established containing the necessary data requirements that will direct the contractors to submit incurred costs that are clearly traceable to the correct CWBS elements where work was actually performed. It should also be noted that correct cost reporting, as well as accurate and meaningful performance data are directly dependent upon a clear, complete, and accurate WBS /CWBS hierarchy.

3.3.2 Compatibility with Internal Management Systems

The WBS is a tool used by NASA, contractor, and other non-NASA performing entities. Management objectives and system capabilities play a significant role in the development of a WBS. NASA and management from all other project performing entities have flexibility in developing a WBS to accommodate their objectives and needs, which include effective and relational management and reporting systems. It is critical to project success for the WBS to be structured and formatted in a manner that is compatible with business/management system capabilities for budgeting, scheduling, cost accounting, time keeping, and performance analysis. The WBS typically serves as the common relational reference point between all these systems to allow for effective interface functionality (see Figure 3-4). The ability for management systems to function as an integrated architecture is a key ingredient to successful management processes.

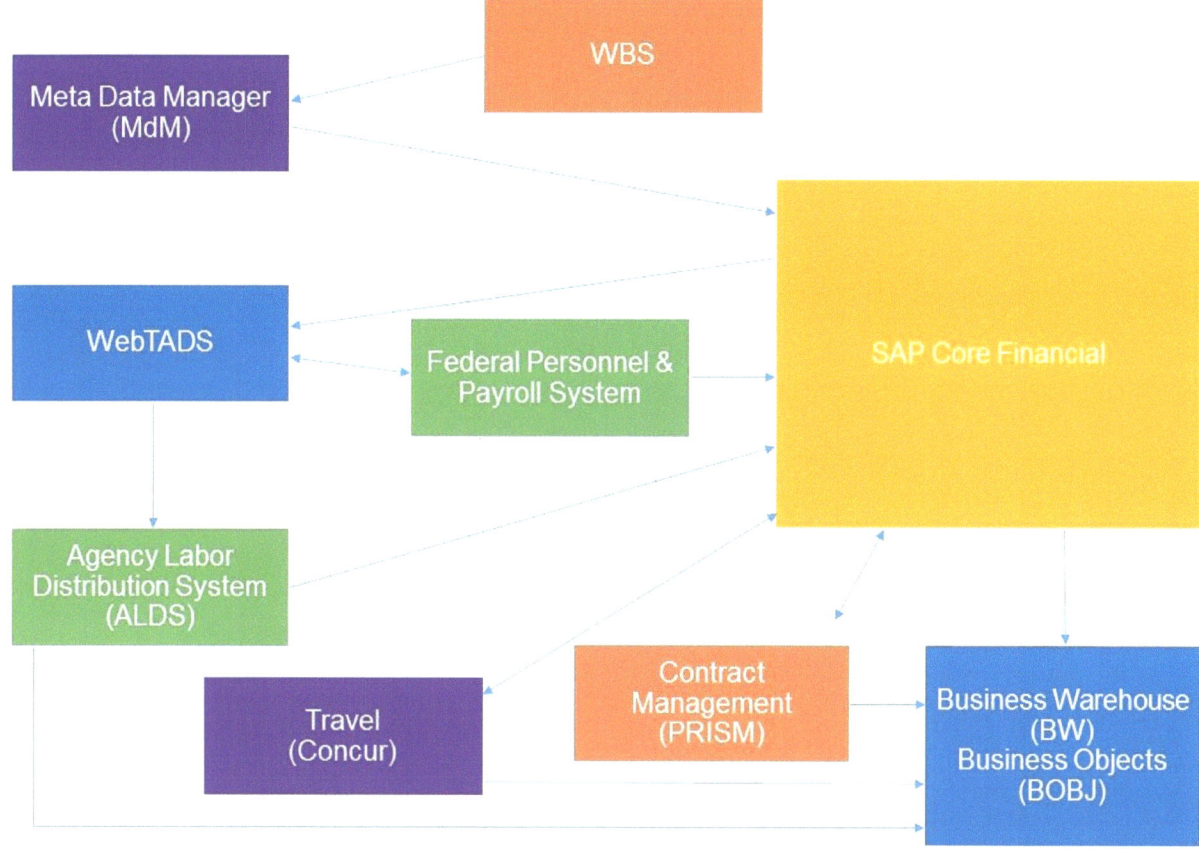

Figure 3-4: WBS Relational Interfaces to NASA Business/Management Systems

3.3.3 Correlation with Other Requirements

NASA and other non-NASA performing entities should give attention to the correlation of the WBS with other requirements of the project, contract, and other applicable partnering agreements. Simply stated, when developing the WBS, do not lose sight of the overall project (or contract/agreement) goals.

The WBS should represent the entire scope of work for the project. This should include all products defined in the project's authorized work scope. This also means that correlation must exist between WBS content definition and all referenced requirement and specification documents. It is not uncommon for the organization performing the systems engineering function to develop a functional block diagram depicting the requirements per the NASA Systems Engineering Handbook (current revision), and then correlate the diagram in developing a product-oriented WBS congruent with the established templates located in NPR 7120.5 (Appendix H), NPR 7120.7/NID 7120.99 Interim Directive (Appendix H), and NPR 7120.8 (Appendix K). While not always necessary, it is useful for ensuring that all desired functionality is included.

Another very useful tool for WBS development is a cross-reference matrix (see Figure 3-5). This matrix typically would have two axes: one horizontal and one vertical. Along one axis a reference to a section or paragraph in the document or documents containing the product definition is listed. Along the other axis the proposed WBS elements are listed. An "X" is then placed in the matrix at each intersection of a requirement and a WBS element. A look across the matrix from the WBS element perspective can quickly identify elements with no corresponding requirement. This condition would indicate work that is not a part of the required scope and should be removed or formally added to the requirements. A look across the matrix from the requirements document or documents perspective will indicate any areas where work scope has not been included in the WBS.

WBS Code	Level 1 2 3 4 5	WBS Element Description	Responsible CAM	SOW Paragraph / Requirements Reference				
123456	X	Flight Project Vehicle	Kerby					
123456.06	X	Spacecraft	Poole					
123456.06.01	X	Spacecraft Management	Poole	X				
123456.06.02	X	Propulsion	Fleming		X			
123456.06.03	X	Architecture	Richards			X		
123456.06.04	X	Avionics	Cucarola				X	
123456.06.05	X	Thermal Control	Terrell					X
123456.06.05.01	X	Passive Thermal Control	Terrell					X
123456.06.05.02	X	Active Thermal Control	Terrell					X
123456.06.05.02.01	X	Active Thermal Control Pumps	Terrell					X
123456.06.05.02.02	X	Active Thermal Control Piping & Valves	Terrell					X
123456.06.05.02.03	X	Active Thermal Control Radiator	Terrell					X
				3.12.01	3.12.02	3.12.03	3.12.04	3.12.05

Work Breakdown Structure Cross-Reference Matrix
Program: Flight Project Vehicle   Contract No: XXXX-XX-XXXX

Figure 3-5: WBS Cross-Reference Matrix

3.3.4 Number of Levels

The hierarchical structure of a WBS is an important consideration. Project work is performed to satisfy technical objectives established for each product or sub-product identified as a WBS element. As each product is subdivided into greater depth within a WBS, each element's technical complexity and resource requirement are reduced. The number of levels and elements in the structure is generally dependent upon the size and complexity of the total effort, the degree of technical uncertainty, organizational structures concerned, and management's judgment of need.

Figure 3-6 illustrates a hierarchical subdivision of work that is product oriented down to a level where work is planned, costs are charged, and management insight is maintained.

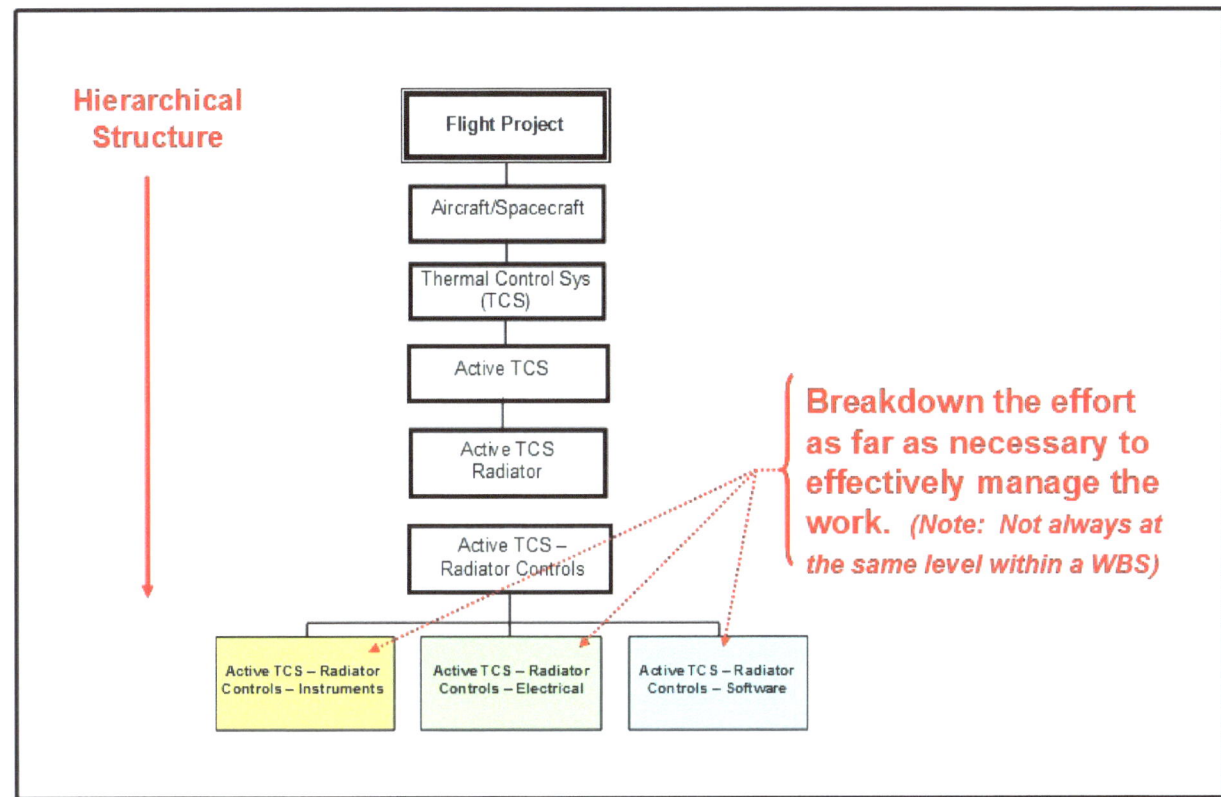

Figure 3-6: WBS Hierarchy Illustration

As the end product is decomposed into smaller sub-products at lower WBS levels, the work effort required by each element can eventually be identified and relatable to an appropriate functional organization. At some level on each WBS branch, management will assign responsibility for technical, schedule, and cost performance. It is at this intersection of WBS element and organization unit that a CA is usually established, work is scheduled, budget is planned, cost is collected, and performance is measured, recorded and controlled. To do this, the technical content and requirements for each work product must be clearly specified and documented. As project work is accomplished, actual completion and technical requirements can be verified.

The WBS level at which a CA is established is primarily a function of the size of the project and the type of product. The responsible organization level is a function of the management span of control and upper management's desire to delegate technical, schedule, and cost responsibility for WBS elements to lower management levels. *It should be understood that all control accounts do not have to be established at the same level within the WBS structure.* A CA, if needed, may be subdivided further into Work Packages (WPs) and Planning Packages (PPs). A WP provides further detail on work content that is considered near-term, while a PP defines far-term work at a summary level. Each product branch within the WBS only needs to be subdivided as far as needed to allow for adequate management, insight, and control.

Control Account, Work Package, and Planning Package are terms typically associated with the management process called Earned Value Management (EVM). Providing an EVM discussion is not the intent of this section. However, due to many NASA projects requiring EVM implementation, and since there is such a close and important relationship between the WBS, CA, WP, and PP, this topic will be addressed briefly. It should be clearly understood by project teams involved in up-front project planning, that the quality and format of their WBS will correlate directly to how well they will be able to implement usable and meaningful EVM practices.

A CA generally consists of one or more WPs and may include one or more PPs. A WP is the unit of work required to complete a specific job such as a test, a report, a design, a set of drawings, fabrication of a piece of hardware, or a service. A planning package is a logical aggregation of work within a CA containing far-term effort which can be identified and budgeted in early baseline planning at a higher level, but cannot yet be adequately defined into a WP. The WBS must be extended at a minimum to the CA level, however, if lower insight is needed to enhance or satisfy management needs, then WPs or PPs may be included within the WBS. If EVM is a project requirement, then careful early coordination with the financial analyst is required to ensure that the WBS contains the needed levels of element detail and that the correct WBS element codes are established as CA(s) and then activated within the Agency's Core Financial System to accept actual costs. This will enable the project to have the capability to gather actual project costs at the necessary level of WBS detail to allow for meaningful performance measurement.

With project content hierarchically structured and defined in the manner described above, it is then possible to satisfy the CA manager with specific data needs. Technical, schedule, and cost information needed to effectively manage an organization's work can be easily supplied by the Center's management control systems by filtering to reflect only the appropriate WBS/CA element information.

Generally, the guidance provided in the above paragraphs will also apply to a contractor's CWBS or any other non-NASA performing entity's WBS development process. As stated earlier in this document, a contractor's extended CWBS will normally begin at level three or four. Additional levels of sub-division may be included prior to starting the CWBS when it is necessary to more clearly communicate all project requirements, or if there is a significant degree of technical risk or complexity. Prime contractors, as well as other non-NASA performing entities should also extend their hierarchical elements in a predominantly product-oriented fashion. The RFP/agreement should instruct all performing entities to extend their CWBS/WBS elements to a level as low as necessary to provide meaningful CA planning, insight, and control.

The following illustration (Figure 3-7) shows how CA(s) fit into the overall project management process.

Control Accounts
(Intersection of WBS and OBS)

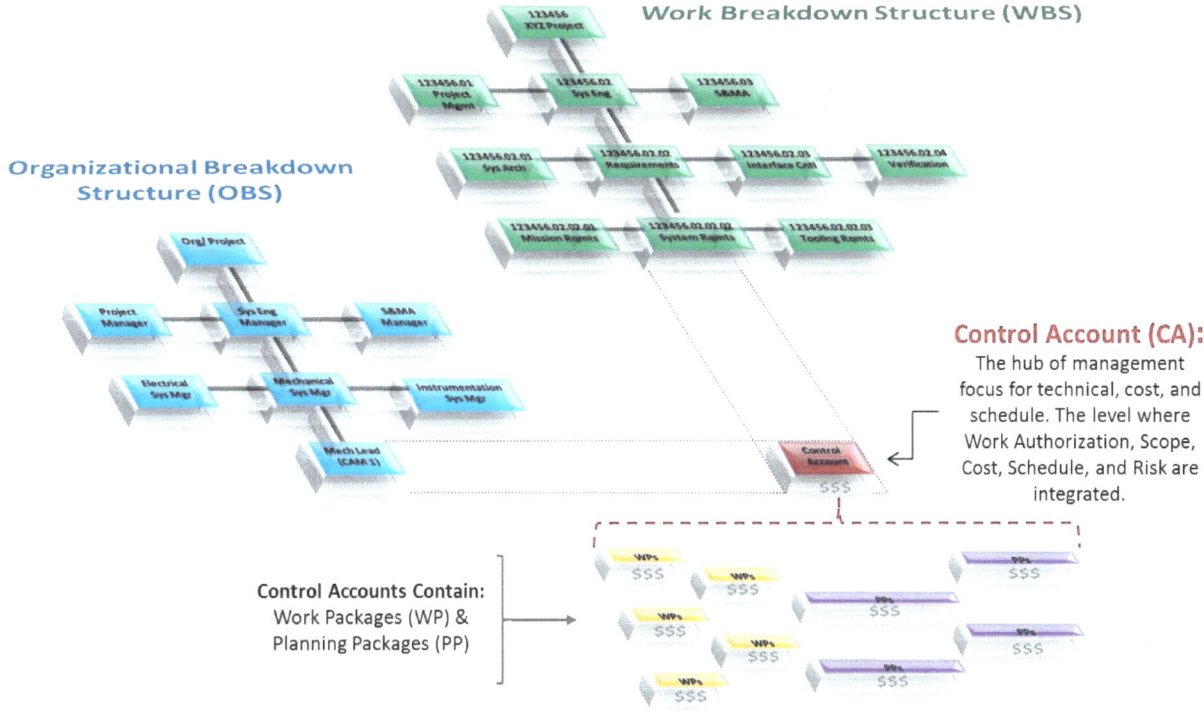

Figure 3-7: Relationships between WBS, OBS, CA, WP, and PP

When identifying CAs, the performing entities must be allowed to establish organizational responsibilities at meaningful and appropriate levels. The CA brings together all aspects of the implementer's management control systems, including technical definition, budgets, estimates, schedules, work assignments, accounting, progress assessment, risk and problem identification, and associated corrective actions.

3.3.5 All Inclusiveness

A WBS should include all NASA in-house, contracted, and other non-NASA performing entity work and products defined in the project scope of work. The WBS should never contain unauthorized work scope.

Major subcontracts, if any, will also need to be included. In some cases, the subcontracted effort may provide for delivery of a single lower-level CWBS element, such as a vendor-fabricated subassembly. In other cases, the subcontract may provide for effort covering several lower-level CWBS elements, such as design for the electronics, communications, and instrumentation systems in a new facility. In either case, the prime contractor's CWBS dictionary (and other management control systems) must be capable of uniquely distinguishing major subcontractors' responsibilities from each other and from the work retained in-house by the prime contractor. This is accomplished at the lower levels of the CWBS.

3.3.6 Change Control

WBS development is an iterative process. While strong efforts should be placed on early and accurate WBS planning, after a baseline is established, WBS revisions will likely still occur from expansion or contraction of project or contract scope, and the movement of a project through its various stages (i.e., engineering, development, and operation). Whenever the baseline WBS is revised, whether it be in-house NASA elements or contractor elements, formal documentation of the revision must be maintained according to the project's configuration and data management process to include the associated change rationale and project manager approval.

Changes may occur as the products of the work effort are more accurately defined or when a revised product structure (resulting from technically different requirements or a more cost effective approach to satisfy the requirements) is used. The WBS and WBS Dictionary should also be revised to reflect changes resulting from contract negotiations.

The extended CWBS is a contractual requirement and specified portions may not be changed without NASA approval. The contract should specify the CWBS elements for which NASA approval is required prior to contractor revision. Usually, the NASA controlled elements are identical to those specified for periodic contractor reporting. Approval of contractor proposed changes to the WBS/CWBS must be made with great care. It is important to always maintain the ability within the overall WBS to identify the interrelationships of each contractor's efforts to the overall project objectives.

A disciplined WBS change control process, as described above for NASA in-house and contracted effort, should also be applied in similar fashion to all other project non-NASA performing entities.

Each project manager is ultimately responsible for establishing and controlling the approved WBS baseline. Project personnel should aid the project manager in this regard by routine monitoring of the WBS codes and associated attribute data contained in the MdM System and other Agency management systems. Monitoring activity should include the review for new, modified, and closed WBS element codes that have been entered into the MdM System without the cognizance of the project manager.

3.4 WBS Development Techniques

The following techniques can be helpful in the development of a project WBS:

    a. Preparing functional requirement block diagrams
    b. Coding WBS elements in a consistent manner
    c. Preparing element tree diagrams
    d. Preparing a WBS dictionary
    e. Using development checklists
    f. Using WBS templates

3.4.1 Preparing Functional Requirement Block Diagrams

A good place to begin developing a preliminary WBS is by listing the functional requirements for the new product or system. If the number of requirements is large, group them into similar categories. Create a parent or end product at the top of your list or graphic. Then, create subordinate major groupings of functional requirements with similar characteristics. Next, subdivide the major groupings into smaller groupings that have natural dividing points. Finally, translate this functional block diagram into a product structure.

In order to translate the functional block diagram into a product structure, begin at the top level with the desired end product that is completely integrated. At the first subordinate level, using the functional block diagram, create a major system block for each related functional block or blocks. The translation may result in a one-to-one, one-to-many, or many-to-one relationship between the functional blocks and the major system blocks. Create subsystem or other subordinate blocks that further define the major system blocks as necessary. Continue this process until the product structure is complete and all functional requirements are accounted for. The result is your initial WBS.

3.4.2 Coding WBS Elements in a Consistent Manner

The NSM System is the internal coding system used by the Agency to define and organize project work content. This same coding system is also used to account for all financial activities associated with funds appropriated by Congress to accomplish project work. When developing a project WBS, the management team coordinates with the financial organization to establish a WBS coding structure that will not only work for technical management of the effort, but also consistently matches the coding structure used by the NASA financial community for cost accounting.

Each WBS element is assigned a unique element code to be used as a reference point for its technical and financial identification throughout the life of the project. A simple decimal coding system that logically indicates the level of an element and related lower-level subordinate elements is applied. A common coding nomenclature facilitates communications among all project participants. Each CWBS, or other non-NASA performing entity WBS coding arrangement, must be traceable to the NASA WBS coding system. The NSM coding structure is illustrated below.

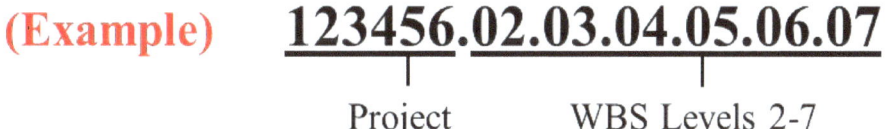

NASA Structure Management (NSM)
WBS Numbering System

- 6 digit project identifier (numeric only)
- 2 digits for remaining levels 2 – 7 (numeric)
- 1 digit for periods between each level 2 - 7

(Example)   **123456.02.03.04.05.06.07**
                   Project   WBS Levels 2-7

★ Center designator is **not** part of the WBS number. Center designator should be placed in an associated attribute field within SAP

Figure 3-8: Agency WBS Numbering System

As illustrated above, each NASA Project (Level 1) is assigned a six-digit numeric code such as 123456. Each sub-level code contains a two-digit numeric code separated from the previous level code by a period. Level 2 codes would appear as 123456.02 and Level 4 codes would look like 123456.02.07.05. A maximum of seven levels is allowed in the WBS coding structure contained in the Agency Core Financial System. The typical lowest level might appear as 123456.02.07.05.02.06.03 containing a maximum of 24 characters including all the periods.

3.4.3 Preparing Element Tree Diagrams

WBS tree diagrams are routinely developed to provide a visual display of a WBS. A pictorial view of the overall WBS aids all project staff in understanding how lower-level project components support and contribute to higher-level components. This type of diagram is often called a "family tree" or a "product tree" diagram.

Examples of WBS tree diagrams are shown below. The first illustration, Figure 3-9, reflects a portion of a Flight Project WBS structure with several key recommended development practices noted. The second illustration, Figure 3-10, portrays a sample portion of a Software Development Project WBS that focuses on software products sub-divided by their functionality.

Partial Project WBS Element Tree Diagram
(Recommended WBS Development Practices Highlighted)

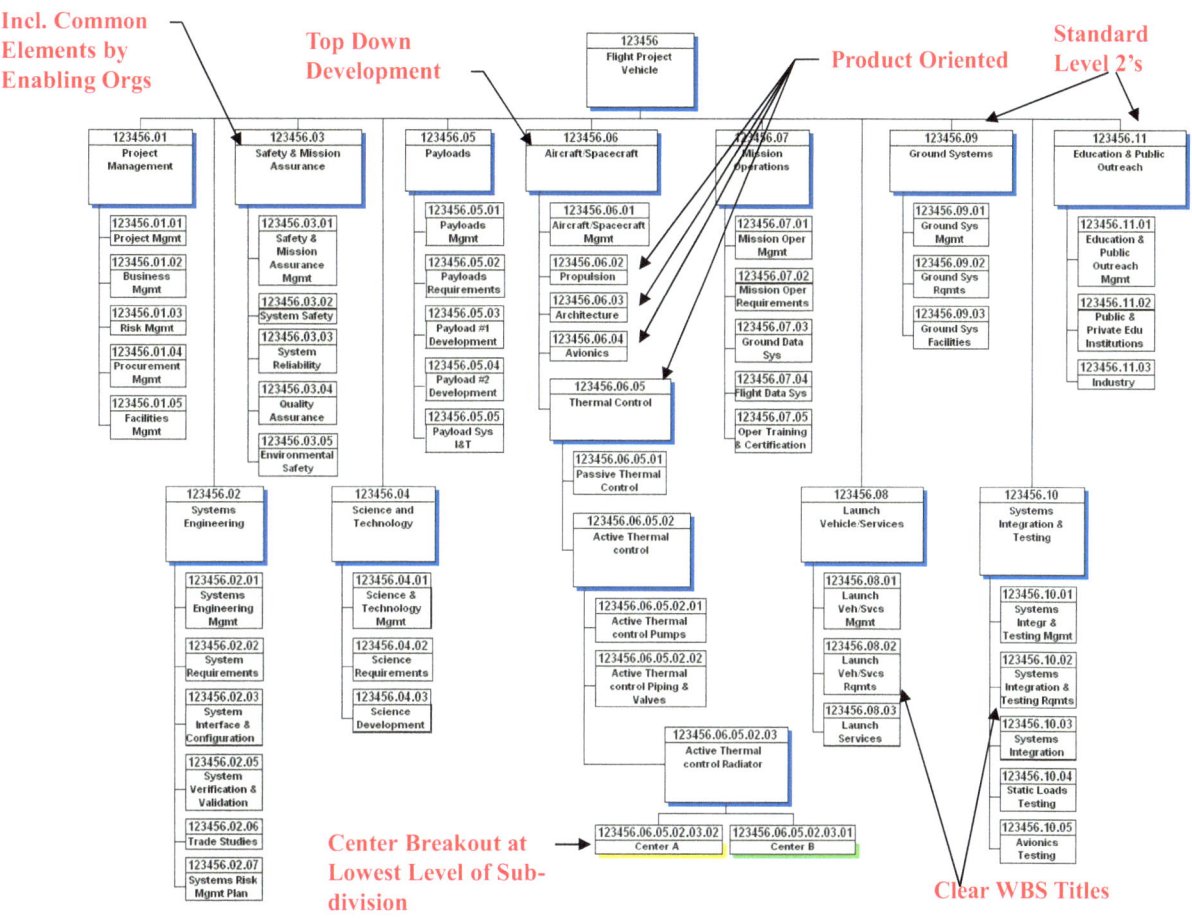

Figure 3-9: Partial WBS Tree Diagram Illustrating Recommended Practices

Sample Software Project WBS

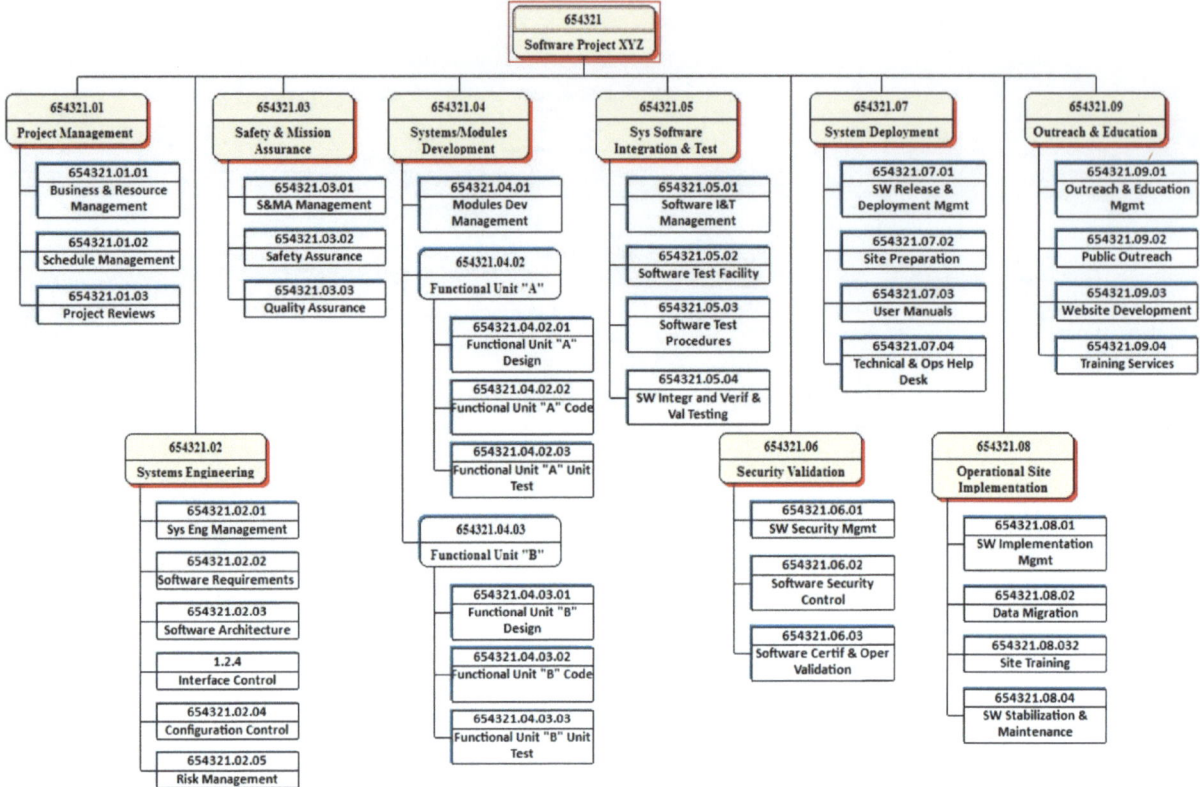

Figure 3-10: Sample Software WBS Illustration

3.4.4 Preparing a WBS Dictionary

A WBS dictionary lists and defines the WBS element contents. It is prepared by the project team for the purposes of ensuring that all work scope has been identified and to eliminate duplication and overlap of work assignments. Content ambiguities should be eliminated with clear statements describing the effort to be completed. The level of descriptive detail needed for each element should be commensurate with the element's hierarchical position in the overall WBS structure. The lower the level of a WBS element, the greater the level of descriptive detail needed. The descriptive detail of element content should clearly identify all element interfaces. The WBS Dictionary is expanded in greater detail for those extended levels established by contractors as the CWBS is developed. Other non-NASA project performing entities will also develop and expand their WBS Dictionary descriptive content in the same fashion as noted above for contractors.

The WBS dictionary element content descriptions should be clear enough to assist in providing a link to the detailed technical documents. The dictionary also contains an index (see Figure 3-11) which lists the WBS elements in indented format to show their hierarchical relationship.

WBS Index

WBS #	WBS Level	WBS Element Title
123456	1	**Project XYZ**
123456.01	2	**Project Management**
123456.01.01	3	Project Management
123456.01.02	3	Resource & Schedule Management
123456.01.03	3	Configuration Management
123456.01.04	3	Project Reviews
123456.02	2	**Systems Engineering**
123456.02.01	3	Sys Engineering Management
123456.02.02	3	Requirements Dev & Verification
123456.02.03	3	System & Mission Analysis
123456.03	2	**Safety & Mission Assurance (SMA)**
123456.03.01	3	S&MA Management
123456.03.02	3	Mission Assurance
123456.03.03	3	Systems Safety
123456.04	2	**Science**
123456.04.01	3	Science Management
123456.04.02	3	Science Requirements
123456.04.03	3	Science Test Procedures
123456.04.04	3	Science Data Analysis
123456.05	2	**Payload**
123456.05.01	3	Payload Management
123456.05.02	3	Payload Electrical Sys
123456.05.03	3	Payload Structures & Mech
123456.05.04	3	Payload Avionics
123456.05.05	3	Payload Thermal Control
123456.05.06	3	Payload Integr &Test
123456.05.07	3	Payload GSE

Figure 3-11: Example - WBS Index Excerpt

Each element definition should include the following:

a. WBS element title
b. WBS element code
c. WBS element content description (including quantities, relevant associated work, and contract end items where applicable)
d. WBS Index
e. Scope definition paragraph number
f. Specification (number and title) associated with the WBS element (if applicable)

g. Date, revision number, revision authorization and approved changes
h. Budget and reporting number (i.e., Charge Code)

Contract WBS Dictionary should include all the above as well as the following:

i. Contract line item associated with the WBS element
j. Contract Identification Number

A WBS dictionary (Figure 3-12) should be a controlled document, and as such, should be maintained as a baseline document with revisions being reviewed and approved according to each project's configuration control requirements for the life of the project.

<u>WBS Dictionary</u>

WBS Element Title: Develop Prototype Code

WBS Element No: 123456.08.05.09.01.03

Parent WBS No: 123456.08.05.09.01

WBS Level: 6

Parent WBS Title: Develop Prototype

Scope Def. Ref: 02.01.01

Project: Project XYZ Software Development

Originator: Samuel L. Kates

WBS Element Description:

The scope of this element includes the development of all necessary software code required to satisfy the functional requirements established for prototype software. This effort will include the identification of all prototype use-case functions, code development for all identified use-case functions, informal use-case testing, integration of all use-case code, and software preparation for full prototype user testing.

Technical Specification Document:

SPEC-SW-AA000765-1

SPEC-SW-AB000767-2

Procurement Packages: None

Related Work Excluded:

1) Associated supervision
2) Prototype requirements validation
3) PP&C planning and control effort

Requirements Doc. No:

NASA SRD-BR549-01

WBS Index: See pages 16-18

Revision No: 01c

Page _15_ of _31_

Revision Date: 11/30/2009

WBS Element No: 123456.08.05.09.01.03

Figure 3-12: WBS Dictionary Example

3.4.5 Using Development Checklists

Checklists are a useful tool to ensure proper WBS development. A WBS checklist helps ensure that all major factors have been considered. Consider the following checklist example (Figure 3-13):

(1) ____Has the correct level 1 and 2 WBS template been used?
(2) ____Are the WBS elements predominantly product-oriented? (Not functional or organizational)
(3) ____Does the overall WBS structure include 100% of the project scope of work? (This should include enabling support and products such as project management, safety & mission assurance, systems engineering)
(4) ____Does the sub-division of WBS elements reflect accurate, logical, and compatible hierarchy of work scope?
(5) ____Does the WBS Dictionary provide complete and explicit content descriptions?
(6) ____Does the WBS sub-divide the project scope of work down to an adequate level of detail to provide for effective resource planning, management insight, and performance measurement?
(7) ____Do the project WBS elements correlate with the following:
   a. Project specification tree
   b. NASA system engineering requirements
   c. Functional design criteria
   d. Technical scope of work
   e. Project Integrated Implementation Plan
   f. Project Integrated Master Schedule (IMS)
   g. NASA internal reporting level requirements

The following checklist items pertain to those WBS elements where contractor or other performing entity responsibilities are involved.

(8) ____Is the contractor's CWBS and/or other performing entity WBS coding compatible with the overall project WBS coding? (Is summarization and roll-up of contractor/other entity data into the appropriate WBS element possible?)
(9) ____Is the contractor's CWBS and/or other performing entity's WBS compatible with the responsibility assignments and management systems?
(10) ____Do the CWBS/other performing entity elements correlate with the following:
   a. Specification tree
   b. End items
   c. Data reporting requirements
   d. Scope of work
   e. Implementation Plan
   f. Integrated Master Schedule?
(11) ____Has the contractor and/or other performing entities defined distinct and logical product-oriented elements down to the level where such definitions are meaningful and necessary for management purposes?
(12) ____Do the CWBS elements encompass 100% of all the project work contracted?
(13) ____Do the CWBS elements encompass all the products of all the work to be performed under subcontract?
(14) ____Does CWBS Dictionary provide complete and explicit content descriptions?
(15) ____Do other performing entity WBS elements encompass 100% of all the work agreed upon?
(16) ____Does the WBS Dictionary provided by other performing entities contain complete and explicit content descriptions?

Figure 3-13: WBS Checklist Example

3.4.6 Using WBS Templates

Since the purpose of a WBS is to divide the project into manageable pieces of work for better planning and control, it only makes sense to add more consistency to this process through the use of standard WBS templates. This standardization not only aids in sub-dividing the work in current projects, but will also make possible the creation of historical data repositories of cost, schedule, and technical information with standard content categories to aid in the planning of future projects. Standard WBS templates are intended to apply to projects, not programs. There are no program WBS standard requirements due to the variance in structure of the Mission Directorates.

As noted earlier, the top two levels of a project WBS are dictated and controlled by the Agency through standard templates. These templates, along with their associated narrative content descriptions, are contained in NPR 7120.5 (Appendix H), NASA Space Flight Program and Project Management Handbook (current revision – section 5.9.1), NPR 7120.7/NID 7120.99 Interim Directive (Appendix H), and NPR 7120.8 (Appendix K). These Agency templates are also found in Appendix C of this document. WBS levels 3 and lower are developed and controlled by project management and, as required, prime contractors that are involved in project implementation.

3.5 Common Development Errors

Developing a properly structured and effective project WBS is a complex process. Increased training in the "best practices" for effective WBS development will help reduce the errors that are commonly found. Examples of common WBS development errors are illustrated below.

3.5.1 Using Unsuitable Former WBS

If a WBS from a prior project is used as a basis for WBS development on a new project or contract, be careful not to perpetuate any mistakes or undesirable features of the earlier WBS.

3.5.2 Non-Product Elements

WBS elements should be predominantly product oriented. For example, Design, Engineering, Manufacturing, Phase A, Pipe Fitters, and Direct Labor are not products and typically should not be used for WBS element sub-divisions. Design, Engineering, and Manufacturing are functions and/or organizations, and Pipe Fitter is a skill type, Direct Labor is a labor category, and Phase A is a time frame. None are appropriate as WBS elements. Rework, Retesting and Refurbishing are additional non-product oriented terms and are not usually recommended as appropriate for WBS element hierarchy.

There is a natural tendency for project managers and contractors to ask, "Where am I in that WBS?" They feel more comfortable if they see specific elements which reflect their functional or organizational areas of responsibility. Since projects many times are functionally organized, the tendency may be for a project team to approve a functionally- or organizationally-oriented WBS. This error should be avoided. The figures below (Figure 3-14 and 3-15) illustrate typical non-product oriented errors.

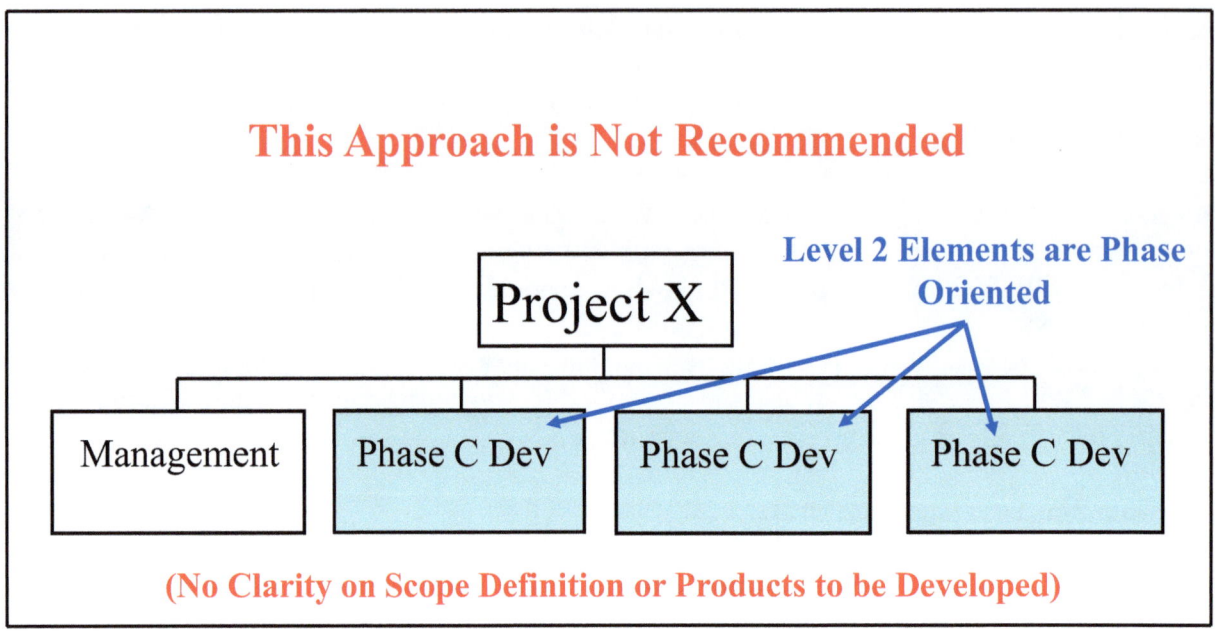

Figure 3-14: Unsuitable Non-Product, Phase-Oriented WBS

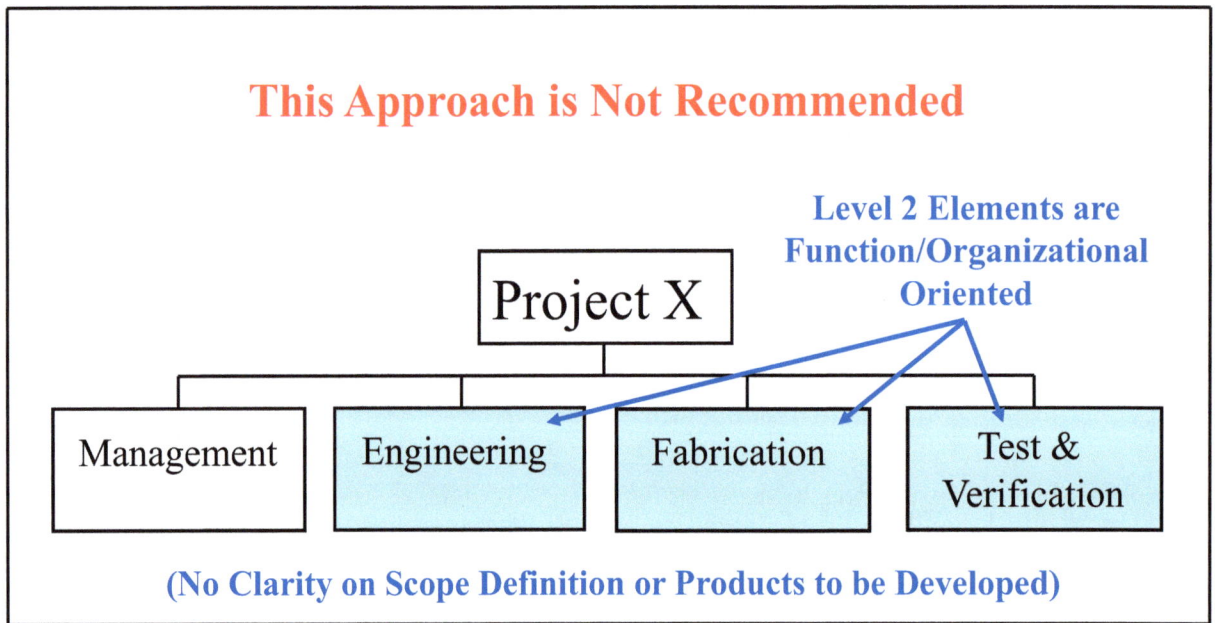

Figure 3-15: Unsuitable Function/Organizational Oriented WBS

3.5.3 Center Breakouts at Inappropriate Levels

NASA projects commonly have involvement from multiple Centers. Current Agency accounting requirements dictate that all Centers with responsibilities for implementing selected portions of work scope within a project must be uniquely identified within the project WBS. Caution must be exercised when including these 'other Center' work elements within the WBS. A common error in developing a project WBS with multiple Center involvement is to insert the Center element at too high a level of the WBS and not at the lower level where work is actually being planned and accomplished. It should be structured in a manner that includes Center-designated elements at the lowest level needed where time is actually charged and costs are collected *(not necessarily the seventh level)*. The reason for this recommendation is to prevent work elements from being fragmented at too high a level. This prevents automated data roll-up and breakdown capabilities that capture the total hardware system development costs and performance for specific system hardware. This error also results in wasted levels of the project WBS so that there may not be enough levels remaining to adequately subdivide the work scope. It should be noted that with only seven WBS levels available within the NSM and financial accounting systems, the project team must use them wisely. The following example (Figure 3-16) illustrates this common error.

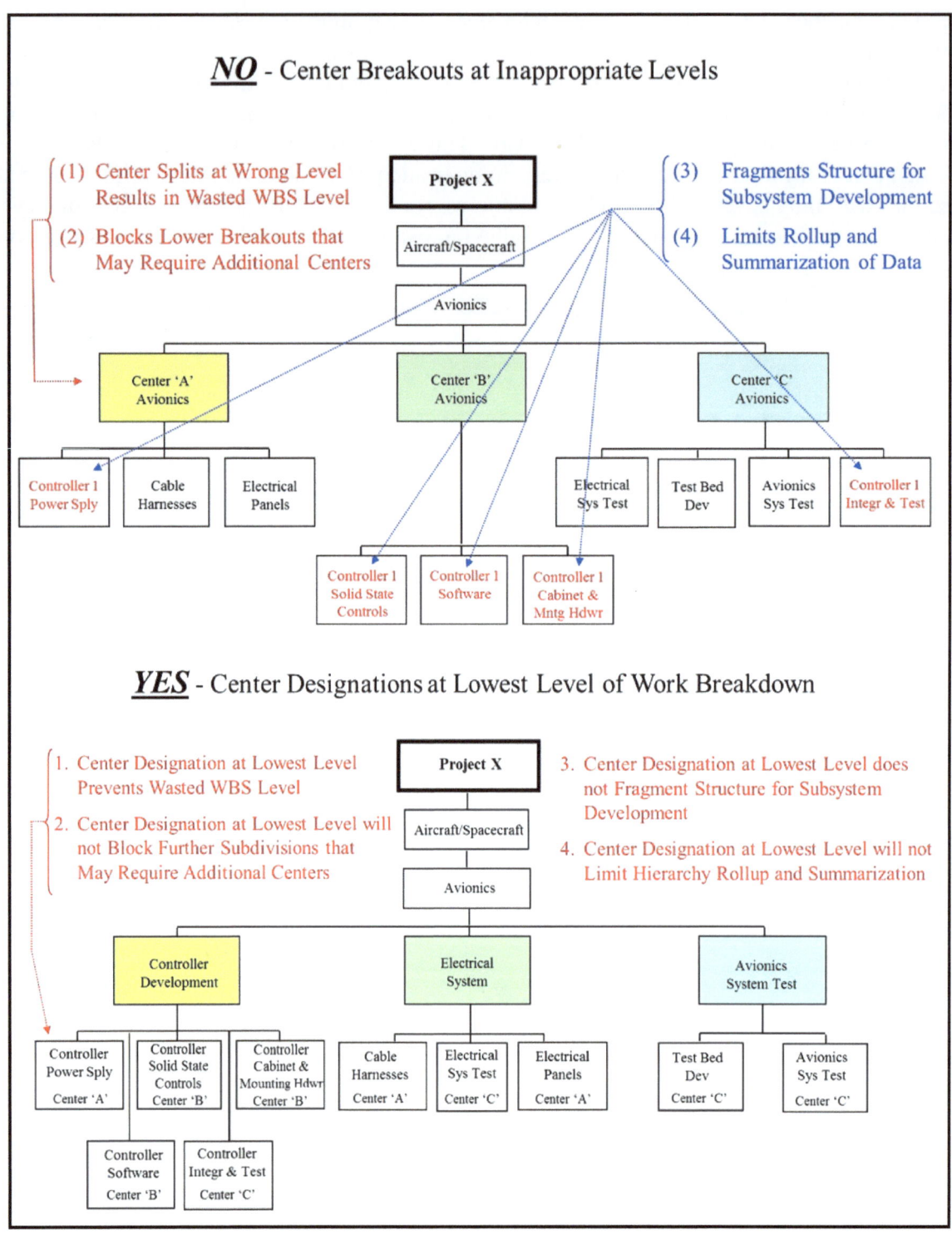

Figure 3-16: Center Breakout Guidance for a WBS

3.5.4 Incorrect Element Hierarchy

A common error in WBS development is sub-dividing the elements in a manner that doesn't correctly reflect the hierarchy of work structure. This error many times is due to not having the right personnel involved in the WBS development effort. A lack of understanding of the total work scope, system hardware, and the implementation processes make it nearly impossible to correctly to build a WBS without the responsible and knowledgeable people involved. The simple example below (Figure 3-17) illustrates an incorrect sub-division of work elements.

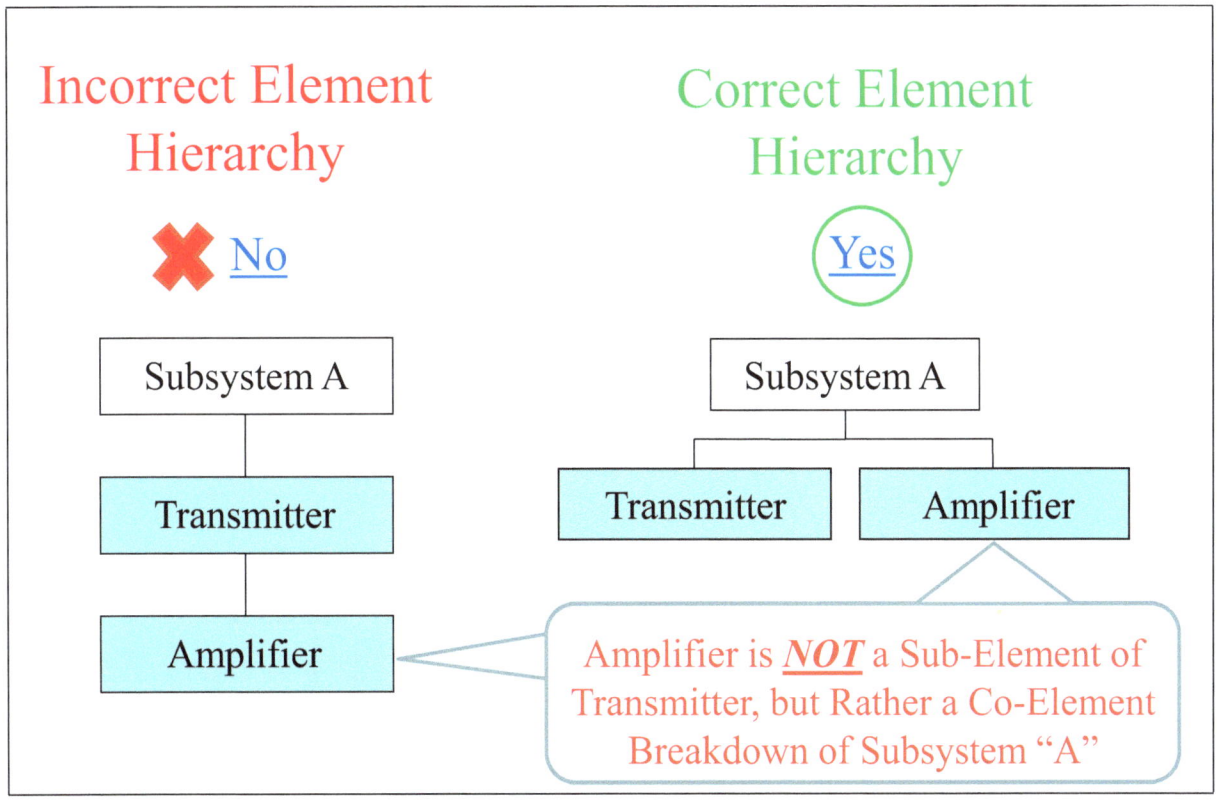

Figure 3-17: Illustration of Incorrect Element Hierarchy

Chapter 4: WBS Uses

The WBS is a project management tool. It provides a framework for specifying the technical aspects of the project by defining the project in terms of hierarchically-related, product-oriented elements for the total project scope of work. The WBS also provides the framework for schedule and budget development. As a common framework for cost, schedule, and technical management, the WBS elements serve as logical summary points for insight and assessment of measuring cost and schedule performance. Figure 4-1, shown below, illustrates how the WBS provides a common reference capability in relating cost, schedule, and technical data.

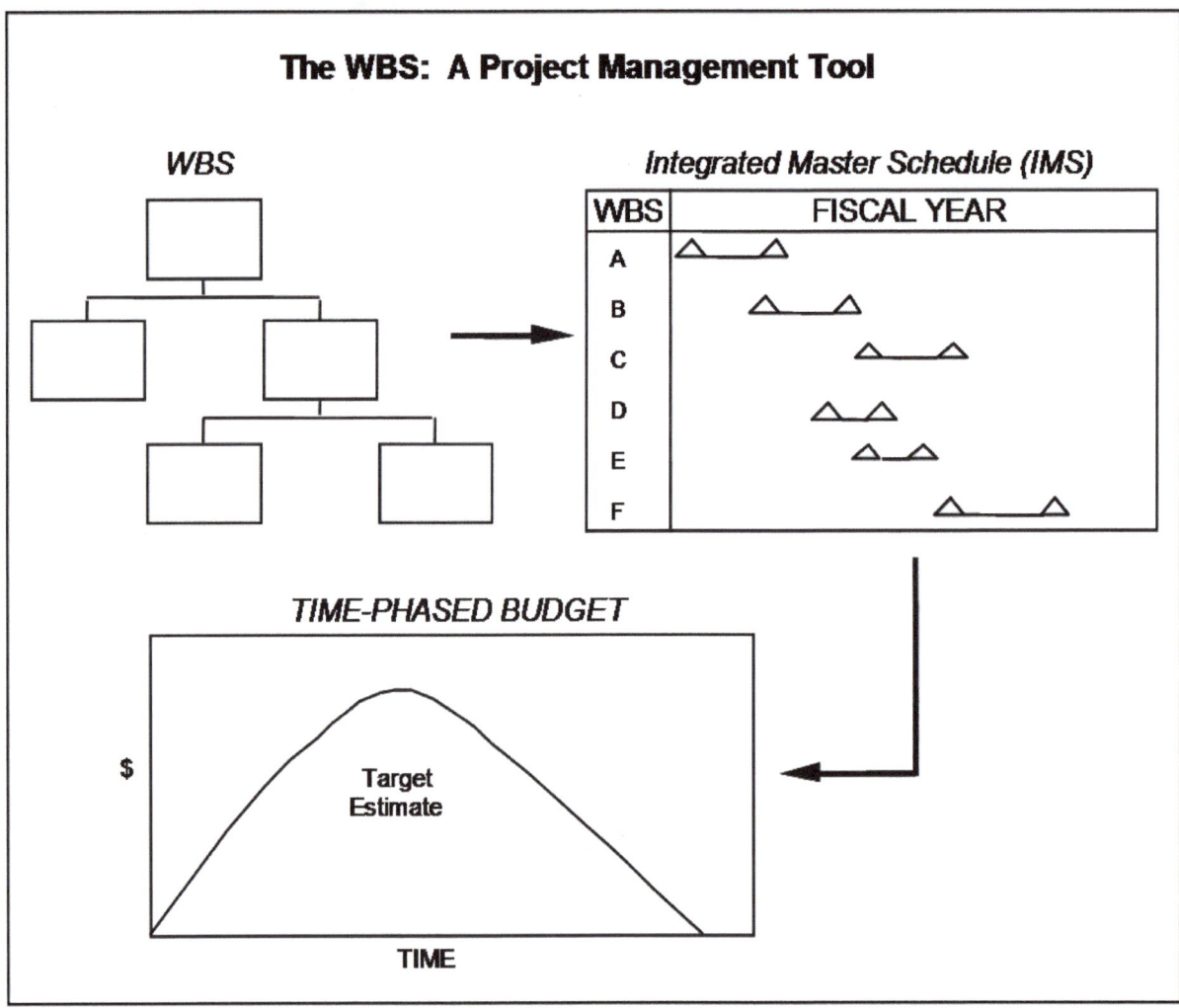

Figure 4-1: The WBS as a Project Management Tool for Integration

The following paragraphs discuss using the WBS for:

  a. Technical Management
  b. Work Identification and Assignment
  c. Schedule Management
  d. Cost Management
  e. Performance Management

f. Risk Management

4.1 Technical Management

The WBS provides the framework for defining the technical objectives, establishing a specification tree, defining configuration items, providing integrated logistic support (ILS), and preparing and executing a test and evaluation plan for a project.

4.1.1 Specification Tree

A specification tree, developed by system engineering, structures the performance parameters for the system or systems being developed. It subdivides the system(s) into its functional constituent elements and identifies the performance objectives of the system(s) and its elements. The performance characteristics are explicitly identified and quantified. The completed specification tree represents a hierarchy of performance requirements for each element of the system for which design responsibility is assigned. Because specifications may not be written for each WBS element, the specification tree may not map to the entire WBS structure.

Technical specification tree development is a sub-element of the standard level two "Systems Engineering" WBS element.

4.1.2 Configuration Management

Configuration Management (CM) includes the process of managing the technical configuration of elements being developed. In establishing the requirement for project configuration management, each project office designates which technical documents and deliverables are subject to CM controls.

A technical document and/or contract deliverable designated for CM is called a configuration item. CM involves defining the baseline configuration for the configuration items, controlling the changes to that baseline, accounting for all approved changes, and verifying that all changes have occurred. The WBS is the framework for designating the configuration items on a project. Thus, the WBS needs to be extended sufficiently to clearly define all elements subject to configuration management.

CM effort should be a sub-element of the standard level two "Systems Engineering" WBS element.

4.1.3 Integrated Logistics Support

Integrated logistic support (ILS) includes all support necessary to assure the effective and economical support of a project, system, or equipment for the project's life cycle. ILS efforts include:

a. Spare and repair parts inventories, warehousing, and control
b. Preventive and scheduled facilities and equipment maintenance
c. Reliability and maintainability data
d. Transportation and handling systems
e. Test equipment
f. Training
g. Related publications

ILS efforts require detailed definitions of the systems or components being supported down to the individual spare part or individual component receiving preventive maintenance. The WBS provides a hierarchical basis for such detailed definition.

ILS effort should be a sub-element of the standard level two "Systems Engineering" WBS element.

4.1.4 Test and Evaluation

Many projects require a formal test and evaluation plan to ensure the acquired element products satisfy the project's objectives as defined by the project technical baselines. Test plans may be developed for individual WBS elements. The effort associated with conducting and evaluating those tests may become part of the work defined for the appropriate WBS element.

Since tests may involve entire systems, parts of systems or individual components, they may not be uniquely identifiable to a single WBS element, but may be included in two or more of the elements going into one higher-level WBS element. Such integrated test work may be planned in conjunction with other work on the higher-level element that incorporates the elements being tested. Lower-level component or sub-assembly test effort may be established as sub-elements under the associated upper-level hardware WBS element for planning and controlling all tests.

Integrated testing of the project's completed, overall level two systems (e.g. payloads, aircraft/spacecraft, launch vehicle/services) should be established as sub-elements of the standard level two "Systems Integration and Testing" WBS element. This includes both procured and NASA developed element products.

4.2 Work Identification and Assignment

People performing work are organized to facilitate effective management, whether the organization is designed along project, functional, or matrix lines. To assign specific work responsibility to a specific organization, the WBS and organizational structure should be integrated with each other (i.e., functional responsibility is established for managing specified work to produce defined products). This integration process results in a Responsibility Assignment Matrix (RAM) which is illustrated below in Figure 4-2. This integration can occur at any level of the WBS and at whatever level of responsibility has been assigned to manage the work. While an interface relationship may exist between the organizational structure and the WBS keep in mind that the organizational structure should not drive how the WBS is structured and sub-divided. The WBS hierarchy should be driven totally by the makeup of the work content and how it can most effectively be sub-divided for management planning, insight, and control.

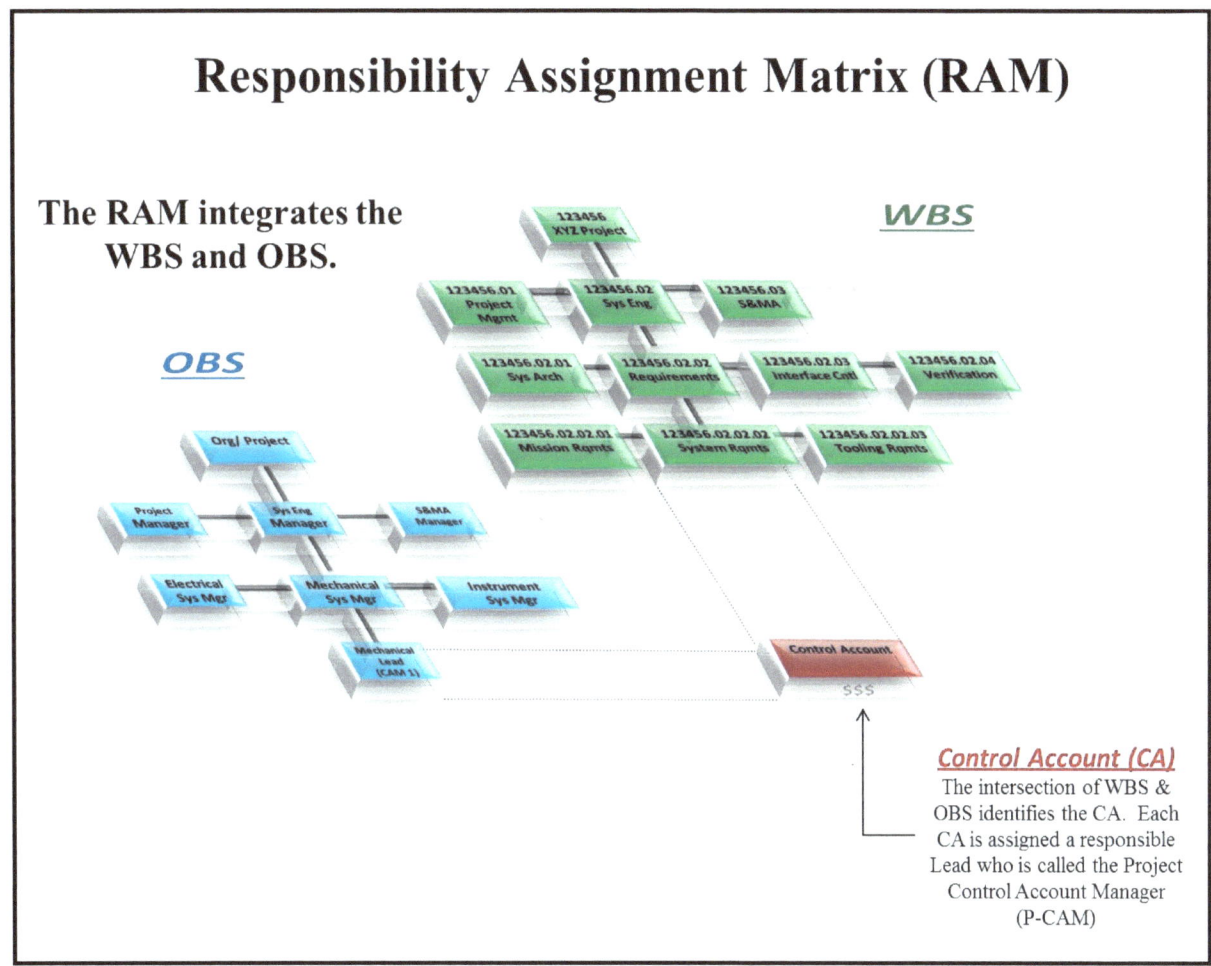

Figure 4-2: Responsibility Assignment Matrix (RAM)

4.3 Schedule Management

The WBS provides a framework for detailed project planning and schedule development. As WBS elements are established and work content is clearly defined in the WBS Dictionary, it is then possible for the project team to determine the tasks (activities) and events (milestones) that are required to successfully complete the project objectives and products. Tasks and events are identified for the effort contained in each lowest-level WBS element. These are then included in project schedules and organized in a format that enables management to determine the proper time-phasing of work. This process should be done using an automated project management scheduling tool that has functional capabilities for task sequencing, slack (float) calculations, task start/finish date generation, critical path determination, constraint identification, and resource loading. As tasks for each of the lowest WBS elements are input into the scheduling tool, various types of associated data are also input for each task, such as description, duration, sequencing, responsibility, constraints, resources, and WBS coding. With all task information in place in the schedule tool for all lowest-level WBS elements, the result is an Integrated Master Schedule (IMS). Because WBS coding is included in the IMS, task networks (logic networks) and other schedule data can be produced at the lowest detail or summarized to various levels of the WBS structure for project management insight, assessment, and control. The NASA Schedule Management Handbook can be referenced for additional scheduling process guidance.

Project schedules are typically available at detailed, intermediate, and summary levels. The level of schedule detail needed is dictated by the scope and complexity of the project work and the needs of management for schedule visibility and control. Schedule levels and management levels need not coincide with WBS levels. Generally there is no requirement for separate schedules for each WBS level. However, if the appropriate WBS element numbers are assigned to each lowest-level task and milestone contained in the IMS, then through data filtering all schedule items can be extracted and formatted by lowest detail or summarized by any WBS level. The lowest level of each WBS element should have at least one task or activity.

Since a product-oriented WBS serves as the framework for schedule development, then resulting project schedules are also product-oriented. This type of schedule allows managers to monitor the schedule baseline for the project's products to ensure that the project objectives are completed on time. Prime contractors must also submit required schedule reporting, with CWBS coding included, to permit government insight and control at the necessary levels of detail.

Project schedule management effort should be established as a sub-element of the standard level two "Project Management" WBS element.

4.4 Cost Management

The WBS is a key tool used for assisting project teams in managing cost. By breaking the total product into successively smaller entities, management can more readily verify that all work identified by the WBS and budgeted for (and later charged against) actually contributes to the project objectives. By using an approved project WBS to plan the total scope of work, it serves as the framework for cost estimating, budgeting, accounting, and control of project costs.

The WBS structure and content definition provides a systematic approach to cost estimating that helps ensure that relevant costs are not omitted. A project estimate that is based on the WBS helps NASA management to ensure that budget development is credible and complete. When WBS elements and the supporting work are scheduled, a solid basis for time-phased budgets is ready-made. The WBS also provides a common framework for tracking the evolution of estimates (e.g., conceptual estimates, preliminary design estimates, and detailed design estimates). The WBS can also provide a framework for life cycle cost analysis.

As periodic project cost estimates are developed, each succeeding estimate is made in an attempt to forecast more accurately the project's total cost. Basically, the estimates may be organized in two ways, by WBS element or by code of accounts. Both support NASA's on-going efforts in preparing budgets and evaluating contractor performance.

It is important to remember that NASA requirements for project management mandate that projects implement a technical WBS which matches exactly the accounting WBS contained in the Agency Core Financial System. Accounting transactions must be coded in such a way that they can be identified to the specific project WBS element which incurred the transaction cost, and to the time period when the transaction occurred. Since the WBS element number also serves as the financial charge code, the accounting system can be programmed to accept charges at levels of the WBS hierarchy where insight is necessary based on project management's authorization.

NASA projects must follow NPR 7120.5, 7120.7 (NID 7120.99), and 7120.8 requirements to use standard level two WBS templates. Because of this, the standard level two elements can be used as categories to accumulate pertinent historical cost data. Such historical data can be used in conjunction with learning curves, regression and other techniques to estimate the cost requirements for similar elements of new projects. Subsequent cost data collected by NASA can be compared to the original estimates to establish their validity, identify trends, and re-estimate future project needs. Contractor and other performing entity data can also be included in this type of historical database if consistent WBS coding is maintained.

Cost management effort should be established as a sub-element of the standard level two "Project Management" WBS element.

4.5 Performance Management

Integrated performance management, also known as Earned Value Management (EVM) begins with a product-oriented WBS that accurately reflects all project work to be accomplished. As stated above, the WBS provides a framework for organizing all technical, schedule, and budget planning. It should be noted that NASA's requirement for the technical and financial WBS to be one and the same will enhance a project team's ability to successfully correlate cost, schedule, and technical planning into a cohesive management baseline. Proper use of the WBS as a common reference for integration of cost, schedule, and technical management accomplishes the performance management objectives of defining and controlling project work.

The integrated performance management process dictates using the same WBS in defining the work scope, in scheduling (time-phasing) the work, and also in establishing budgets that correlate appropriately. The illustration shown below in Figure 4-3 depicts in a simple overview how the Performance Measurement Baseline (PMB) is derived from a resource-loaded schedule. This figure shows how the time-phased accumulation of resources that have been applied to the schedule provide a performance profile that is used for measuring project performance. This baseline is used not only for measuring performance accomplishment but also for deriving trend data for future project cost and schedule projections.

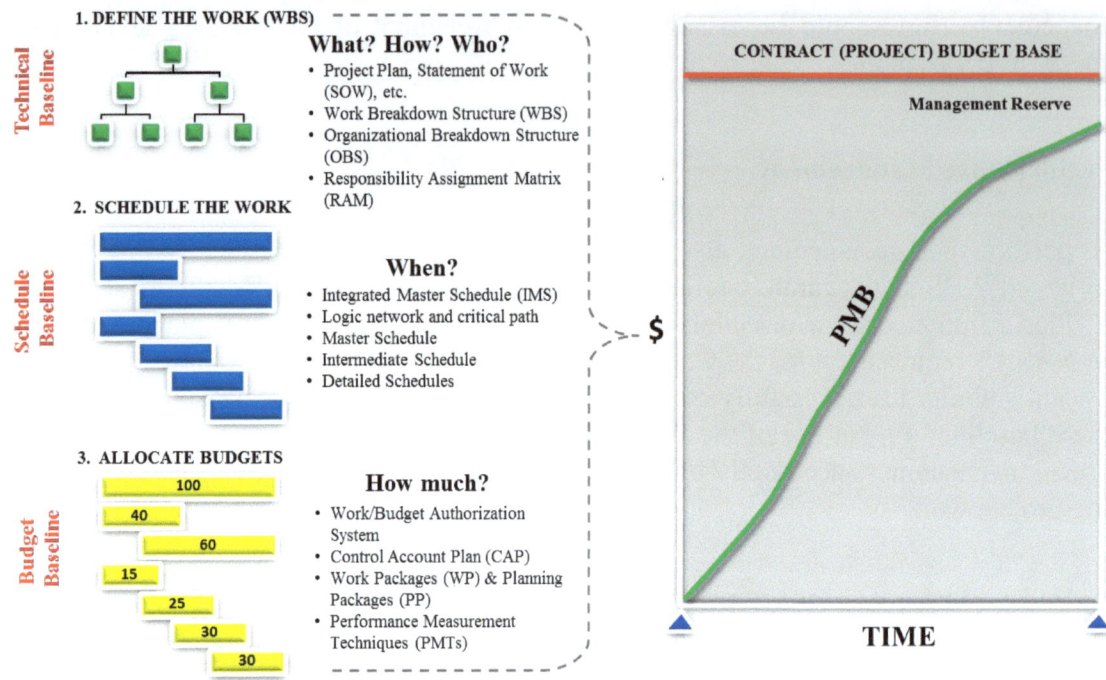

Figure 4-3: WBS and the Development of the Performance Measurement Baseline (PMB)

In addition, the WBS is used to accumulate performance data and associated variances. This allows NASA Project Control Account Managers (P-CAMs) and contractors to evaluate progress in terms of performance for their specific project responsibilities. There is no need for separate performance assessments to be made at levels above the CA because the WBS facilitates the summarization of data for successively higher levels of management insight and oversight. By using the WBS, variances can then be traced to their source areas.

Performance management effort should be established as a sub-element of the standard level two "Project Management" WBS element.

4.6 Risk Management

Risks represent potential impacts to successful achievement of project goals and objectives. Project risks typically fall into various defined categories such as technical, programmatic (cost, schedule, resources), etc. A risk can also exist in any element of work content and at any time during the project life cycle. Since a WBS serves as the common reference for all project cost, schedule, and technical data, it only makes sense that it should also play an important role in managing risks (see Figure 4-4). All risks identified during project implementation can typically be correlated to effort associated with one or more WBS elements. As risks are logged into the risk management system, the WBS element number will be a key data point for further tracking and mitigation planning. Risk mitigation or the lack of risk mitigation will many times impact cost and/or schedules for affected WBS elements. Control of risk impacts is a critical management emphasis and as such will involve on-going mapping to the WBS.

The WBS Provides a Common Reference Point for Documenting, Managing, Tracking, and Communicating Program/Project Risks

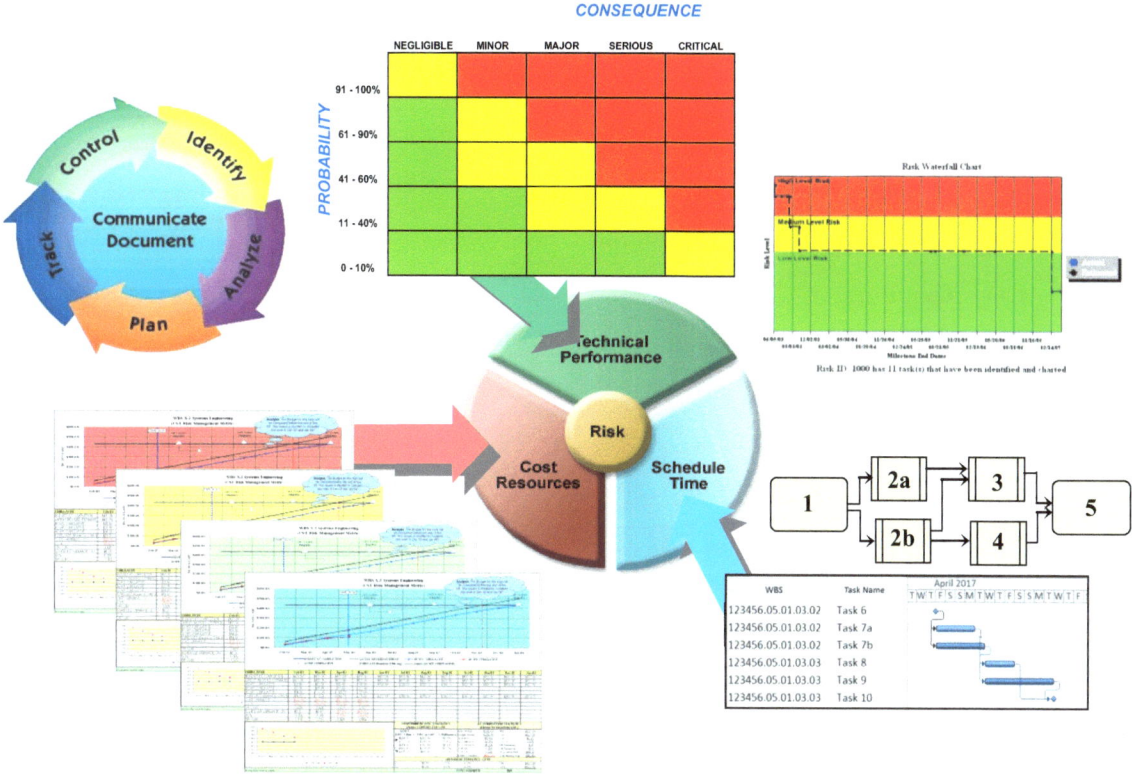

Figure 4-4: WBS Serves as a Common Reference Point in Risk Management

Risk management effort within a project should be established as a sub-element of the standard level two "Systems Engineering" WBS element.

APPENDIX A: Acronym Listing

ALDS	Agency Labor Distribution System
ATP	Authority to Proceed
BOBJ	Business Objects
BW	Business Warehouse
CA	Control Account
CI	Configuration Item
CAM	Control Account Manager
CM	Configuration Management
CWBS	Contract Work Breakdown Structure
DRD	Data Requirements Document
EIA	Electronic Industries Alliance
EPO	Education and Public Outreach
EVM	Earned Value Management
GSE	Ground Support Equipment
HQ	NASA Headquarters
ICD	Interface Control Document
ILS	Integrated Logistic Support
IMS	Integrated Master Schedule
MdM	Metadata Manager System
MIL	Military
NASA	National Aeronautics and Space Administration
NFS	NASA FAR Supplement
NID	NASA Interim Directive
NPD	NASA Policy Directive
NPR	NASA Procedural Requirements
NSM	NASA Structure Management
OBS	Organizational Breakdown Structure
PCA	Program Commitment Agreement
P-CAM	Project Control Account Manager
PMB	Performance Measurement Baseline
PMI	Project Management Institute
PP	Planning Package
RAM	Responsibility Assignment Matrix
R&T	Research and Technology
RFP	Request for Proposal

RTOPs	Research and Technology Objectives and Plans Summaries
SAP	Systems, Applications, Products in Data Processing
S&MA	Safety and Management Assurance
SE&I	Systems Engineering and Integration
SOW	Statement of Work
STI	Science and Technical Information
TCS	Thermal Control System
TD	Technology Development
V&V	Verification and Validation
WBS	Work Breakdown Structure
WP	Work Package

APPENDIX B: Glossary of Terms

Configuration Item (CI) - An aggregation of hardware/computer programs or any of its discrete portions, which satisfies an end-use function and is designated by NASA for configuration.

Contract Work Breakdown Structure (CWBS) - A work breakdown structure of the products or services to be furnished under contract. It is comprised of selected Project WBS elements specified in the contractual document and the contractor's lower level extensions of those elements.

Control Account (CA) – A documented scope of technical, cost, and schedule objectives within a project corresponding to a WBS element that has a responsible organizational element or individual identified. The control account is represented in a Responsibility Assignment Matrix (RAM) as the intersection of the WBS and the Organizational Breakdown Structure (OBS).

End Item - A final combination of end products components, parts, or materials which is ready for its intended use; an item of software or documentation that is deliverable to a user or customer.

High Risk Item - An item which involves technological, manufacturing or other state-of-the-art advances or considerations, and program/project management designates as requiring special attention. It is critical from the standpoint of achieving program objectives, reliability, maintainability, safety, quality assurance or other such factors.

Integrated Master Schedule (IMS) - An integrated schedule developed by logically networking all detailed program/project activities. The highest level schedule is the Master Schedule supported by Intermediate Level Schedules and by lowest level detail schedules.

Metadata Manager (MdM) System - A web-based application that controls the creation, maintenance, and archiving of the NASA Structure Management (NSM) structure and the NSM Coding System codes.

Performance Measurement Baseline (PMB) - The time-phased budget plan against which performance is measured. It is formed by the budgets assigned to scheduled control account and the applicable indirect budgets. For future effort, not planned to the control account level, the performance measurement baseline also includes budgets assigned to higher level WBS elements and undistributed budgets. It equals the total allocated budget less management reserve.

Program - A strategic investment by a Mission Directorate (or Mission Support Office) that has defined goals, objectives, architecture, funding level, and a management structure that supports one or more projects.

Project - A specific investment identified in a Program Plan having defined goals, objectives, requirements, life-cycle cost, a beginning, and an end.

Project Plan - A detailed plan which, when formally approved, sets forth the agreement between a program manager and project managers, and defines the guidelines and constraints under which the project will be executed.

Request for Proposal (RFP) - A solicitation used in negotiated acquisition to communicate government requirements to prospective contractors and solicit proposals.

Statement of Work (SOW) – A document containing a narrative description of the total work scope for a project or contract.

Specification Tree - A graphic portrayal arranged to illustrate interrelationships of hardware and/or software performance/design requirements specifications; normally, this portrayal is in the form of a "family tree" subdivision of the specifications with each lower-level specification applicable to a hardware/software item which is part of a higher-level item.

Subsystem - A functional entity within a system. The name given for the next level of breakdown under a system.

System - One of the principal functioning entities comprising the project hardware within a project or flight mission. Ordinarily, a system is the first major subdivision of project work.

Work Breakdown Structure (WBS) – A product-oriented hierarchical division of the hardware, software, services, and other work tasks that organizes, displays, and defines the products to be developed and/or produced and relates the elements of the work to be accomplished to each other and the end product(s). The WBS should be accompanied by a text document referred to as a WBS Dictionary that describes the work content each element of the WBS in detail.

WBS Dictionary - A document that describes the work content of each WBS element, in product-oriented terms, and relates each element to the respective, progressively higher levels of the structure, as well as to the Statement of Work.

WBS Element - Any block or unique entry in a work breakdown structure regardless of level.

WBS Levels - The arrangement or configuration of a WBS which establishes the hierarchy of projects to programs, systems to projects, subsystems to systems, etc.

Work Package (WP) – Detailed jobs, or material items, identified by the implementer for accomplishing work required to complete the project/contract. A work package has the following characteristics:
    a) It represents units of work at levels where work is performed.
    b) It is clearly distinguished from all other work packages.
    c) It is assigned to a single organizational element.
    d) It has scheduled start and completion dates and, as applicable, interim milestones, which are representative of physical accomplishment.
    e) It has a budget or assigned value expressed in terms of dollars, man-hours, or other measurable units.
    f) Its duration is limited to a relatively short span of time or it is subdivided by discrete value milestones to facilitate the objective measurement of work performed or it is level-of-effort.
    g) It is integrated with detailed engineering, manufacturing, or other schedules.

APPENDIX C: Standard Project WBS Level 2 Templates and WBS Dictionary Content Descriptions

A. SPACE FLIGHT PROJECT WBS Standard Level 2 Elements (see NPR 7120.5, Appendix H):

Standard Level 2 WBS elements for space flight projects are shown in the figure below. The standard WBS template assumes a typical spacecraft flight development project with relatively minor ground or mission operations elements. For major launch or mission operations ground-development activities which are viewed as projects unto themselves, the WBS may be modified. For example, the spacecraft element may be changed to reflect the ground project major deliverable (such as a facility). The elements such as payload, launch vehicle/services, ground system(s), and mission operations (system) that are not applicable may be deleted. (Appendix E shows an example of a hardware contract work breakdown structure that a hardware-development prime contractor might use.)

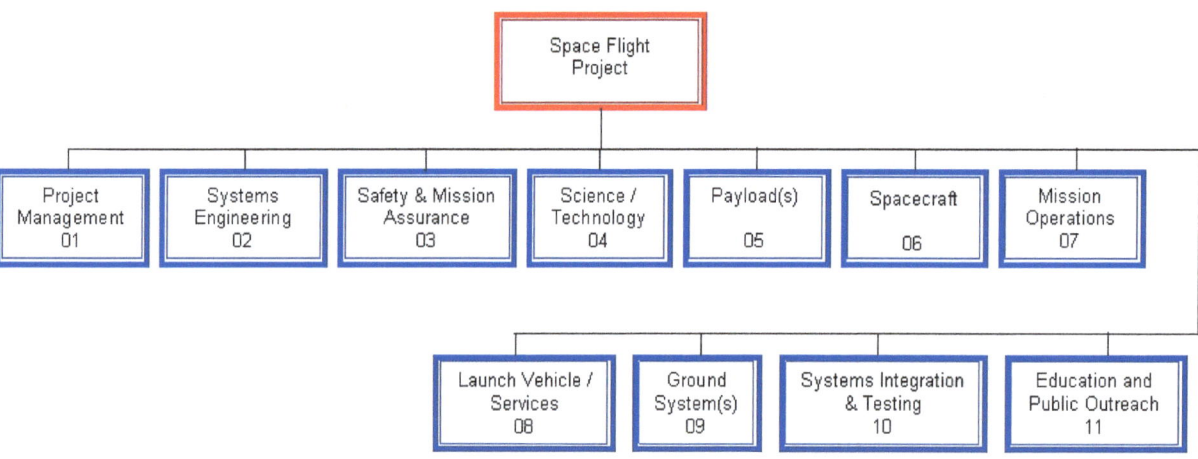

The following is the Space Flight Project Standard WBS Dictionary:

Element 01 - Project Management: The business and administrative planning, organizing, directing, coordinating, analyzing, controlling, and approval processes used to accomplish overall project objectives, which are not associated with specific hardware or software elements. This element includes project reviews and documentation, non-project owned facilities, and project reserves. It excludes costs associated with technical planning and management and delivery of engineering, hardware and software.

Element 02 - Systems Engineering: The technical and management efforts of directing and controlling an integrated engineering effort for the project. This element includes the efforts to define the project space flight vehicle(s) and ground system, conducting trade studies, the integrated planning and control of the technical program efforts of design engineering, software engineering, specialty engineering, system architecture development and integrated test planning, system requirements writing, configuration control, technical oversight, control and monitoring of the technical program, and risk management activities. Documentation products include requirements documents, Interface Control Documents (ICD), Risk Management Plan, and master Verification and Validation (V&V) plan. Excludes any design engineering costs.

Element 03 - Safety and Mission Assurance: The technical and management efforts of directing and controlling the safety and mission assurance elements of the project. This element includes design, development, review, and verification of practices and procedures and mission success criteria intended to assure that the delivered spacecraft, ground systems, mission operations, and payload(s) meet performance requirements and function for their intended lifetimes. This element excludes mission and product assurance efforts directed at partners and subcontractors other than a review/oversight function.

Element 04 - Science / Technology: This element includes the managing, directing, and controlling of the science investigation aspects, as well as leading, managing, and performing the technology demonstration elements of the Project. The costs incurred to cover the Principal Investigator, Project Scientist, science team members, and equivalent personnel for technology demonstrations are included. Specific responsibilities include defining the science or demonstration requirements; ensuring the integration of these requirements with the payloads, spacecraft, ground systems, and mission operations; providing the algorithms for data processing and analyses; and performing data analysis and archiving. This element excludes hardware and software for onboard science investigative instruments/payloads.

Element 05 - Payload: This element includes the special-purpose equipment and normal equipment, e.g., Ground Support Equipment (GSE), integral to the spacecraft. This includes managing, and implementing the hardware and software payloads that perform the scientific experimental and data-gathering functions on board the spacecraft, as well as the technology demonstration for the mission.

Element 06 - Spacecraft(s): The spacecraft that serves as the platform for carrying payload(s), instrument(s), humans, and other mission-oriented equipment in space to the mission destination(s) to achieve the mission objectives. The spacecraft may be a single spacecraft or multiple spacecraft/modules (i.e., cruise stage, orbiter, lander, or rover modules). Each spacecraft/module of the system includes the following subsystems, as appropriate: Crew, Power, Command & Data Handling, Telecommunications, Mechanical, Thermal, Propulsion, Guidance Navigation and Control, Wiring Harness, and Flight Software. This element also includes all design, development, production, assembly, test efforts, and associated GSE to deliver the completed system for integration with the launch vehicle and payload. This element does not include integration and test with payloads and other project systems.

Element 07 - Mission Operations System: The management of the development and implementation of personnel, procedures, documentation, and training required to conduct mission operations. This element includes tracking, commanding, receiving/processing telemetry, analyses of system status, trajectory analysis, orbit determination, maneuver analysis, target body orbit/ephemeris updates, and disposal of remaining end-of-mission resources. The same WBS structure is used for Phase E Mission Operation Systems but with inactive elements defined as "not applicable." However, different accounts must be used for Phase E due to NASA cost reporting requirements. This element does not include integration and test with the other project systems.

Element 08 - Launch Vehicle / Services: The management and implementation of activities required to place the spacecraft directly into its operational environment, or on a trajectory towards its intended target. This element includes launch vehicle, launch vehicle integration, launch operations, any other associated launch services, and associated ground support equipment. This element does not include the integration and test with the other project systems.

Element 09 - Ground System(s): The complex of equipment, hardware, software, networks, and mission-unique facilities required to conduct mission operations of the spacecraft systems and payloads. This complex includes the computers, communications, operating systems, and networking equipment needed to interconnect and host the Mission Operations software. This element includes the design, development, implementation, integration, test, and the associated support equipment of the ground system, including the hardware and software needed for processing, archiving, and distributing telemetry and radiometric data and for commanding the spacecraft. Also includes the use and maintenance of the project test beds and project-owned facilities. This element does not include integration and test with the other project systems and conducting mission operations.

Element 10 - Systems Integration and Testing: This element includes the hardware, software, procedures, and project-owned facilities required to perform the integration and testing of the project's systems, payloads, spacecraft, launch vehicle/services, and mission operations.

Element 11 - Education and Public Outreach: Provide for the education and public outreach (EPO) responsibilities of NASA's missions, projects, and programs in alignment with the Strategic Plan for Education. This element includes management and coordinated activities, formal education, informal education, public outreach, media support, and website development.

B. INFORMATION TECHNOLOGY AND INSTITUTIONAL INFRASTRUCTURE PROJECT WBS Standard Level 2 Elements (see NPR 7120.7 (NID 7120.99 Interim Directive), Appendix H):

The IT Standard WBS provides the needed flexibility for the wide variety of IT projects including Software Development, Hardware Acquisition/Deployment, Service/Process Change etc.

To align with the agency existing project standards, the Standard WBS for IT projects will follow the structure listed below:

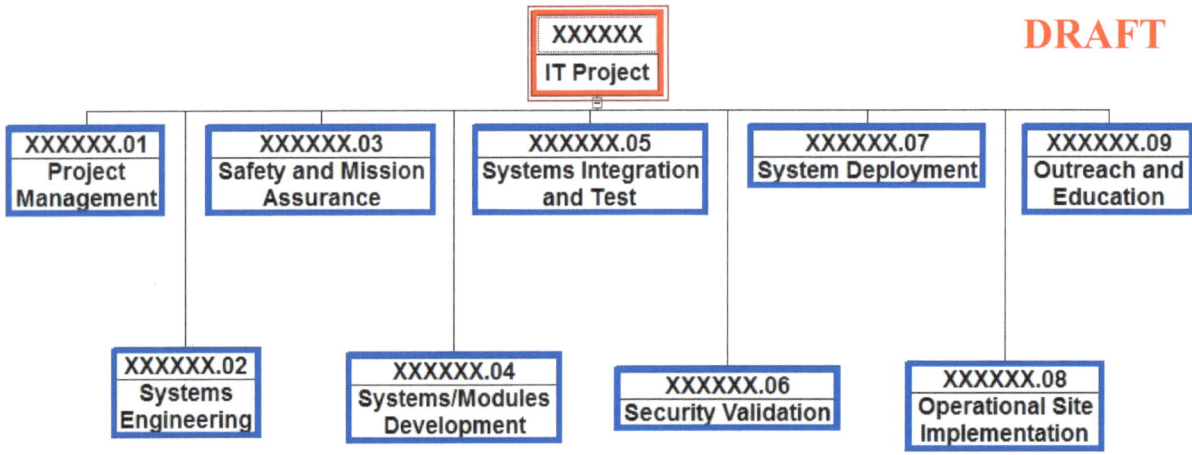

The following is the Information Technology Standard WBS Dictionary (from NPR 7120.7, (NID 7120.99 Interim Directive Appendix H):

Element XXXXXX.01 - Project Management: All activities associated with business and administrative planning, organizing, directing, coordinating, analyzing, controlling, status reporting, and approval processes used to accomplish overall project objectives, which are not associated with specific hardware or software elements. This element includes project KDP reviews and system engineering reviews. It also includes all activities associated with awarding and managing project-related contracts, including technical support contracts. Activities associated with planning and finalizing staffing requirements, facilities, administrative and standard operating procedures are also included. It excludes costs associated with technical planning and management and costs associated with delivering specific engineering, hardware and software products.

Element XXXXXX.02 - Systems Engineering: The technical and management efforts of directing and controlling an integrated engineering effort for the project. This element includes the efforts to define the project IT system(s) and operations delivery processes, conducting trade studies, the integrated planning and control of the technical program efforts of design engineering, software engineering, specialty engineering, system architecture development and integrated test planning, system requirements writing, configuration control, technical oversight, control and monitoring of the technical program. Documentation products include requirements and preliminary design specifications, interface control documents (ICDs), and the system engineering review presentations.

Element XXXXXX.03 – Safety and Mission Assurance: The technical and management efforts of directing and controlling the safety and mission assurance elements of the project. This element includes design, development, review, and verification of practices and procedures and mission success criteria intended to assure that the delivered product meets performance requirements and function for its intended lifetime. This element excludes product assurance efforts directed at partners and subcontractors other than a review/oversight function, and the direct costs of environmental testing. Full compliance with applicable sections of NPD 8700.1 is required for each project

Element XXXXXX.04 – System/Modules Development All activities associated with designing, developing and/or procuring hardware and software configuration items, developing prototypes at the development facility and the resulting integration, unit testing, assembly and checkout. The detailed design is completed for each subsystem and a composite detailed design specifies the relationship of each subsystem or module.

Element XXXXXX.05 - System Integration, Verification and Validation: All activities associated with testing, analyzing and evaluating in order to verify and validate that products meet specifications, satisfy requirements and are operationally suitable and effective. This element includes system integration and user acceptance testing.

Element XXXXXX.06 – Security Validation: The technical and management efforts of directing and controlling the security related elements of the project. This element includes security assessment, security categorization, Certification and Accreditation activities, scanning activities and final validation for the capability to move into full operation.

Element XXXXXX.07– System Deployment: All activities required to deploy the system/subsystems/hardware, etc into the production environment using the ITIL ―Service Transition‖ process for ―Release and Deployment Management‖. The WBS includes all site preparation activities, e.g., new power, installation of equipment racks, and cabling. Activities to ensure documentation of standard operating procedures, planning for final security scans and any other activities required to transition operations to a ―production‖ facility are also included in this WBS. Deployment activities for this WBS address both hardware and software or service/process deployment at either a single production facility or at multiple sites. This WBS would be used for activities to complete the approach for multi-site deployment.

Element XXXXX.08 – Operational Site Implementations: All activities required to implement (move into daily operation) at a specific site. The WBS would be used to track site resource requirements and scheduled site-specific activities for each site. Any site specific activities and resources would be tracked in this WBS. Examples of site specific activities would be the site readiness review, a data migration plan and a transition plan if migrating from one process/system to another.

Element XXXXXX.09 – Outreach and Training: All activities required to provide for the education and public outreach (EPO) responsibilities of NASA's IT projects in alignment with the Strategic Plan for Education. This WBS includes management and coordinated activities, formal education, informal education, public outreach, media support, and website development (if used for outreach activities). Includes all activities associated with designing, developing, and delivering training services, aids

C1. TECHNOLOGY DEVELOPMENT (TD) PROJECT WBS Standard Level 2 Elements (from NPR 7120.8, Appendix K) - formerly Advanced Technology Development Project:

Standard Level 2 WBS elements for TD projects are shown below. The standard WBS template assumes a typical TD project with no flight elements. For TD projects, additional WBS elements may be added horizontally (i.e., at Level 2) as long as their content does not fit into the content of any existing standard WBS elements. The elements that are not applicable do not need to be used (entered into MdM). For TD projects managed as Space Flight Projects in accordance with NPR 7120.5, *NASA Space Flight Program and Project Management Requirements*, the Space Flight Project WBS will be used.

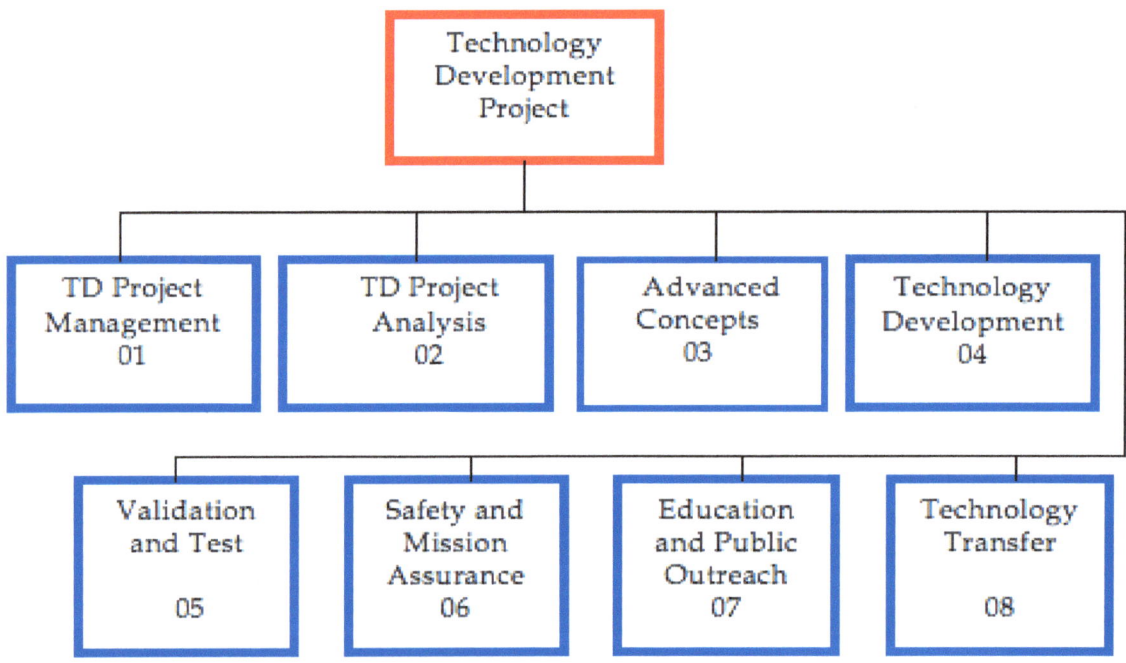

The following is the Technology Development (TD) Project Standard WBS Dictionary (from NPR 7120.8, Appendix K):

Element 01 -- TD Project Management: The business and administrative planning, organizing, directing, coordinating, controlling, and approval processes used to accomplish overall project objectives, which are not associated with specific hardware or software elements. This element includes project reviews and documentation, non-project owned facilities, and project reserves. It excludes costs associated with technical planning and management and costs associated with delivering specific engineering, hardware, and software products.

Element 02 -- TD Project Analysis: System and Portfolio Analysis includes the process of developing qualitative and quantitative understanding of key technical issues and drivers, including current limitations and challenges. These analyses are the foundations that support development and assessment of: goals, requirements, scope, risk/feasibilities, costs, design, integration, and operations. Based on the systems analysis, the process further identifies and tracks the investment trade spaces and supports the project's ability to optimize its resources to maximize the return on investment within acceptable risk exposure, budget, schedule, and performance requirements.

Element 03 -- Advanced Concepts: This WBS element encompasses low-level studies (possibly including laboratory experiments) intended to explore the feasibility of new ideas or approaches to accomplish programmatic or technical objectives. Often they are inspired by new scientific or technical breakthroughs that open up new avenues for technological investigation. Advanced concept activities can be an integral part of a specific technology project, or they can be separate technology activities focused on broad topics. Typically, advanced concept activities are tied to long-term objects, though this is not required, and due to their speculative nature, are generally considered high-risk. Also, they tend to be managed in a looser manner than technology development. Advanced concept studies are a principal means for identifying promising new opportunities for high payoff technology development.

Element 04 -- Technology Development: This WBS element encompasses the execution phase of implementing a TD project plan. Typically, most of the lower level (Level 3 and below) WBS elements associated with achieving the technical objectives (e.g., performance metrics) of the project are contained within this element. TD projects span the gap from advanced concepts to engineering or advanced development (Technology Readiness Level (TRL) 2/3 to TRL 6/7) and can begin and end anywhere within this range. Once a project has started, technology development will include the continual assessment of progress, redistribution of resources and schedule updates necessary to meet key milestones within budget and schedule. When the intended outcome cannot be met within plan, re-planning at the project level falls within Project Management. The primary deliverables, including technical progress reports and documenting technical accomplishments, are part of this WBS element.

Element 05 -- Validation and Test: This element provides for specific activities to test and validate products of technology development when those activities represent a critical aspect of the overall technology development plan. Typically, a separate test and validation element will be part of the project that intends to achieve TRL 6/7. The element encompasses the development hardware/software test validation articles; development or acquisition of special test or validation equipment; scheduling and staffing facilities or ranges; as well as the development and execution of the test or validation plan. Often the full context of test and validation may not be known at the beginning of the project and will be developed as required. In some cases, if extensive use of large-scale facilities is required or the test/validation takes on the characteristics of a space flight project, it will be conducted as a flight project (e.g., space-flight project) within this WBS element or transferred to a space flight project in accordance with NPR 7120.5, *NASA Space Flight Program and Project Management Requirements.*

Element 06 -- Safety and Mission Assurance: Provides for directing and controlling the safety and mission assurance elements of the project. This element includes design, development, review, and verification of practices and procedures and success criteria intended to assure that the delivered product meets performance requirements and function for their intended lifetimes. This element excludes mission and product assurance efforts at partners/subcontractors other than a review/oversight function.

Element 07 -- Education and Public Outreach: Provide for Education and Public Outreach responsibilities of NASA's missions, projects, and programs in alignment with the Strategic Plan for Education, including formal and informal education, public outreach, media support, and Website development.

Element 08 -- Technology Transfer: This WBS focuses principally on three types of activities: (1) Transferring knowledge and technology development products developed within the project to non-NASA entities that are not part of the project either as direct participants or as direct beneficiaries of the project, (2) Acting on behalf of the project to identify and transfer into the project knowledge or technology from sources not directly participating in the project or benefiting from the project, and (3) Providing supporting expertise to transfer knowledge and technology products between NASA and non-NASA entities, including project participants. Typically, technology transfer requires special expertise not associated with any specific project or required to accomplish primary technical objectives/ milestones/metrics. Also, technology transfer activities are often an integral part of the overall project plan but not necessarily an integral part of the technology development process. As such, it is often appropriate to fund, implement, and manage technology transfer as a distinct element of a TD project.

C2. RESEARCH & TECHNOLOGY (R&T) PORTFOLIO PROJECTS WBS Standard Level 2 Elements (from NPR 7120.8, Appendix K): - formerly Basic and Applied Research Portfolio WBS

Standard Level 2 WBS elements for R&T Portfolio Projects are shown below. The template is used to specify the integrated budget within the R&T Portfolio Project Plan. For R&T Portfolio Projects, additional WBS elements may be added horizontally (i.e., at Level 2) as long as their content does not fit into any existing WBS elements. The elements that are not applicable need not be used. For R&T Portfolio Projects managed as Space Flight Projects in accordance with NPR 7120.5, the Space Flight Systems Project WBS will be used.

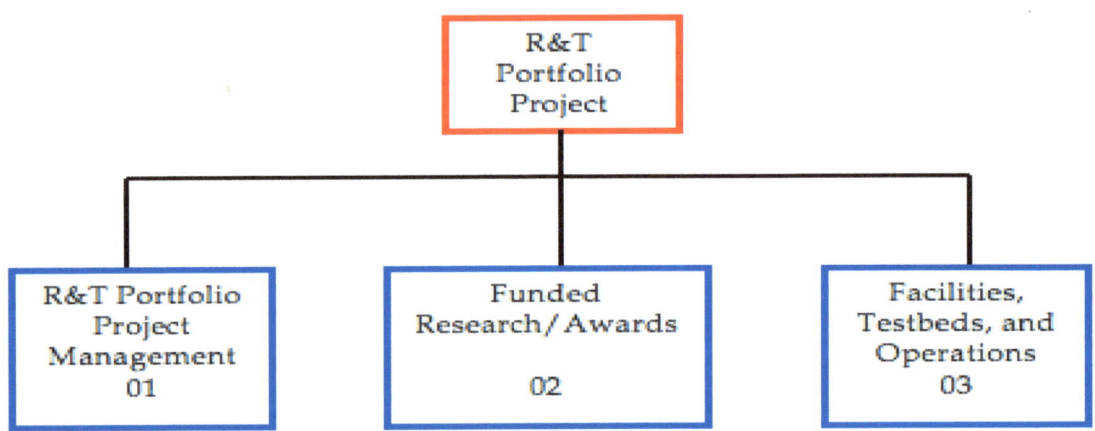

Following is the R&T Portfolio Project Standard WBS Dictionary (from NPR 7120.8, Appendix K, formerly Basic and Applied Research Portfolio Standard WBS Dictionary):

Element 01 -- R&T Portfolio Project Management: The management activity is paid for with R&T Portfolio Project dollars. This includes salaries and travel (e.g., civil service personnel, Intergovernmental Personnel Act assignees, detailees), peer review management (contractor support, travel, and honoraria), and meetings and conferences.

Element 02 -- Funded Research/Awards: The awards from either competed or directed elements of the R&T Portfolio Project. This includes but is not limited to grants, Independent Assessment Teams, Research and Technology Objectives and Plans Summaries (RTOPs), and contracts. This element may also include awards for Education and Public Outreach.

Element 03 -- Facilities, Test beds, and Operations: The non-award infrastructure costs that support the R&T Portfolio Projects.

APPENDIX D: Standard Data Requirements Document (DRD)

DATA REQUIREMENTS DESCRIPTION (DRD)

1. **DPD NO.**: XXX   **ISSUE**: Standard
2. **DRD NO.: STD/MA-WBS**
3. **DATA TYPE**: 2
4. **DATE REVISED**:
5. **PAGE**: 1/2

6. **TITLE**: Work Breakdown Structure (WBS) and WBS Dictionary

7. **DESCRIPTION/USE**: To establish a product-oriented framework for reporting program cost, schedule, and technical performance. To provide a basis for uniform planning, reporting status, program visibility, and assignment of responsibilities.

8. **OPR**: CS40   9. **DM**:

10. **DISTRIBUTION**: Per Contracting Officer's letter

11. **INITIAL SUBMISSION**: Draft with proposal

12. **SUBMISSION FREQUENCY**: 30 days after Authority to Proceed (ATP), update as required. Revised pages shall be submitted 10 calendar days after contract WBS changes (following Government approval).

13. **REMARKS**: Reference is made to NPD 7120.4 (Current Revision), *NASA Engineering and Program/Project Management Policy*, and NPR 7120.5 (Current Revision), *NASA Space Flight Program and Project Management Requirements*, NPR 7120.7 (Current Revision), *NASA Information Technology and Institutional Infrastructure Program and Project Management Requirements*, NPR 7120.8 (Current Revision), *NASA Research and Technology Program and Project Management Requirements* and *NASA Work Breakdown Structure (WBS) Handbook* (Current Revision). These documents shall be used as guides in the preparation of the WBS and the WBS dictionary.

14. **INTERRELATIONSHIP**:

15. **DATA PREPARATION INFORMATION**:
15.1 **SCOPE**: The Work Breakdown Structure (WBS) establishes a product-oriented logical subdivision of hardware, software, services, facilities, etc., that make up the total project scope of work. The WBS Dictionary provides a narrative description of the tasks and effort to be performed in each WBS element.

15.2 **APPLICABLE DOCUMENTS**: None

15.3 **CONTENTS**:   The WBS and WBS Dictionary are two distinct project documents used for defining the approved project scope of work. The contents of each document are detailed in the following paragraphs:
   a. WBS - A logical, hierarchical display of the subdivision of all project work to be completed. The WBS shall include the approved element title and element number.
   b. WBS Dictionary - The WBS dictionary shall describe and document the work content of every WBS element and relevant efforts associated with each element (e.g., design, development, manufacturing). The WBS dictionary shall be arranged in the same order as the contract WBS. The WBS dictionary shall include the following for each WBS element:
      1. WBS element title.
      2. WBS element code.
      3. WBS element content description (including quantities, relevant associated work, and contract end items where applicable).
      4. WBS Index.
      5. SOW paragraph number.
      6. Specification (number and title) associated with the WBS element (if applicable).
      7. Contract line item associated with the WBS element.
      8. Date, revision number, revision authorization and approved changes.

DRD Continuation Sheet

TITLE: Work Breakdown Structure (WBS) and WBS Dictionary	DRD NO.: **STD/MA-WBS**
DATA TYPE: 2	PAGE: 2/2

15. **DATA PREPARATION INFORMATION (CONTINUED):**
    9. Contract Identification Number.
    10. Budget and reporting number (i.e., Charge Code).

15.4 **FORMAT**: The WBS shall be in a chart format showing element relationships, arranged in the same order as the WBS provided in the Request for Proposal. The WBS Dictionary shall be ordered in consonance with the WBS index and shall reference each WBS element by its identifier and name.

15.5 **MAINTENANCE**: Changes shall be incorporated by complete reissue.

*Instructions:

Applicability Instructions: This deliverable is a requirement for all projects and may be tailored on an as-needed basis to meet unusual and/or specific project needs. Tailoring of this DRD shall be coordinated with the responsible OPR.

Preparation Information: Draft contract WBS and Dictionary shall be prepared in response to the Government Request for Proposal (RFP) for Government approval/disapproval. The contractor shall expand the WBS down to the level at which the work will be accomplished. Formal contract WBS shall be in accordance with contractor/Government agreement reached during negotiations.

Statement of Work Words: The contractor shall prepare and submit a Work Breakdown Structure (WBS) and WBS Dictionary in accordance with DRD STD/MA-WBS.

NOTE: The instructions on this DRD are not a part of the DRD and should not be included in a DPD.

APPENDIX E: Contractor CWBS Example

Following is an example of a hardware contract work breakdown structure. This example is from the Ares Project and reflects the prime contractor CWBS for the First Stage hardware development.

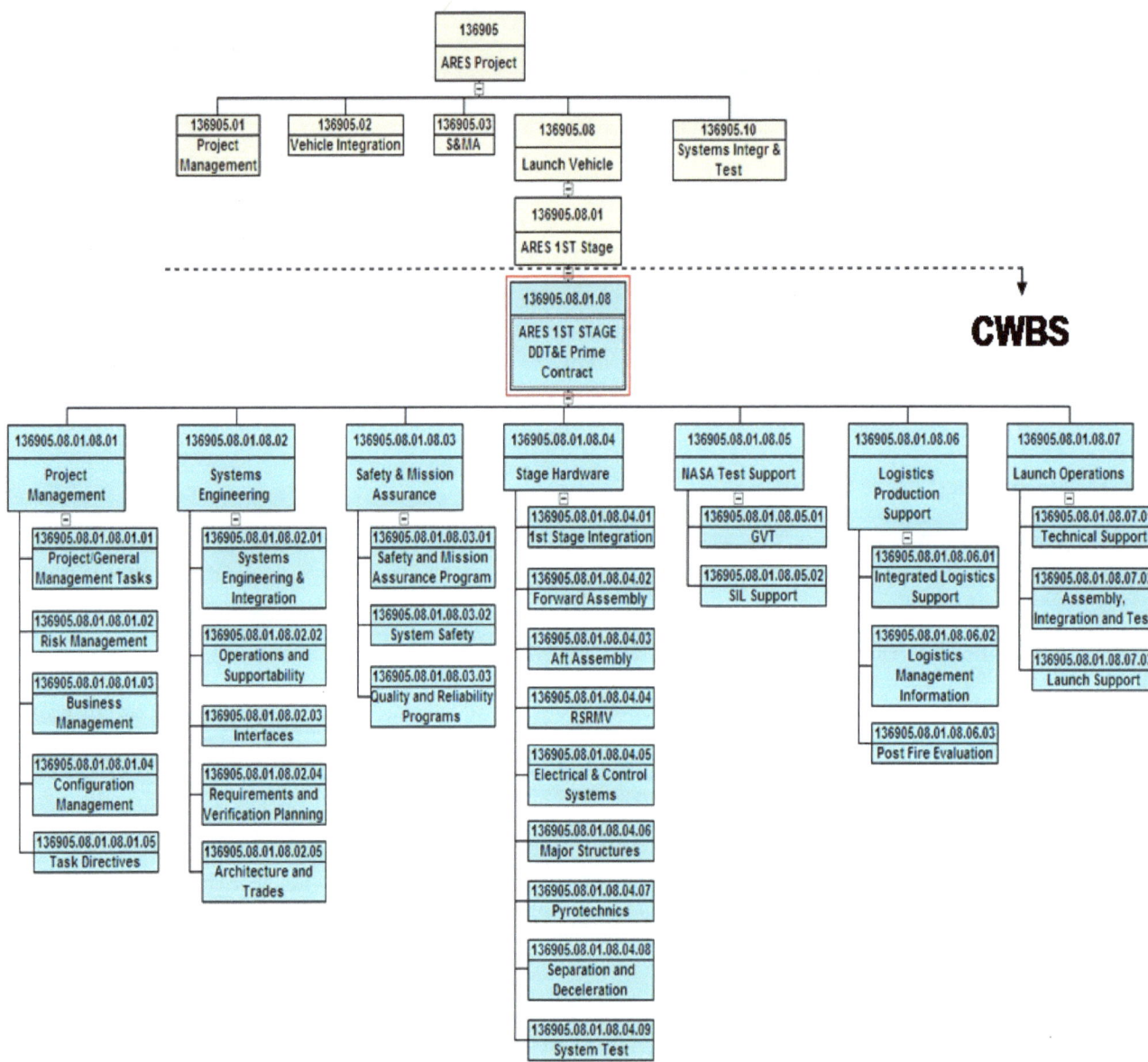

National Aeronautics and Space Administration
NASA Headquarters
Washington, D.C. 20546